Advances in Discrete Applied Mathematics and Graph Theory

Advances in Discrete Applied Mathematics and Graph Theory

Editors

Janez Žerovnik
Darja Rupnik Poklukar

MDPI • Basel • Beijing • Wuhan • Barcelona • Belgrade • Manchester • Tokyo • Cluj • Tianjin

Editors
Janez Žerovnik
University of Ljubljana
Slovenia

Darja Rupnik Poklukar
University of Ljubljana
Slovenia

Editorial Office
MDPI
St. Alban-Anlage 66
4052 Basel, Switzerland

This is a reprint of articles from the Special Issue published online in the open access journal *Mathematics* (ISSN 2227-7390) (available at: http://www.mdpi.com).

For citation purposes, cite each article independently as indicated on the article page online and as indicated below:

LastName, A.A.; LastName, B.B.; LastName, C.C. Article Title. *Journal Name* **Year**, *Volume Number*, Page Range.

ISBN 978-3-0365-4435-9 (Hbk)
ISBN 978-3-0365-4436-6 (PDF)

© 2022 by the authors. Articles in this book are Open Access and distributed under the Creative Commons Attribution (CC BY) license, which allows users to download, copy and build upon published articles, as long as the author and publisher are properly credited, which ensures maximum dissemination and a wider impact of our publications.

The book as a whole is distributed by MDPI under the terms and conditions of the Creative Commons license CC BY-NC-ND.

Contents

Preface to "Advances in Discrete Applied Mathematics and Graph Theory" vii

Xinyue Liu, Huiqin Jiang, Pu Wu and Zehui Shao
Total Roman {3}-Domination: The Complexity and Linear-Time Algorithm for Trees
Reprinted from: *Mathematics* **2021**, *9*, 293, doi:10.3390/math9030293 1

Yaser Rowshan, Mostafa Gholami and Stanford Shateyi
The Size, Multipartite Ramsey Numbers for nK_2 Versus Path–Path and Cycle
Reprinted from: *Mathematics* **2021**, *9*, 764, doi:10.3390/math9070764 9

Paul Bosch, Edil D. Molina, José M. Rodríguez, and José M. Sigarreta
Inequalities on the Generalized ABC Index
Reprinted from: *Mathematics* **2021**, *9*, 1151, doi:10.3390/math9101151 21

Martin Bača, Andrea Semaničová-Feňovčíková and Tao-Ming Wang
Local Antimagic Chromatic Number for Copies of Graphs
Reprinted from: *Mathematics* **2021**, *9*, 1230, doi:10.3390/math9111230 39

Ángel Juárez Morales, Gerardo Reyna Hernández, Jesús Romero Valencia, Omar Rosario Cayetano
Free Cells in Hyperspaces of Graphs
Reprinted from: *Mathematics* **2021**, *9*, 1627, doi:10.3390/math9141627 51

Kiki A. Sugeng, Denny R. Silaban, Martin Bača and Andrea Semaničová-Feňovčíková
Local Inclusive Distance Vertex Irregular Graphs
Reprinted from: *Mathematics* **2021**, *9*, 1673, doi:10.3390/math9141673 63

Abel Cabrera Martínez, Juan C. Hernández-Gómez and José M. Sigarreta
On the Quasi-Total Roman Domination Number of Graphs
Reprinted from: *Mathematics* **2021**, *9*, 2823, doi:10.3390/math9212823 75

Darja Rupnik Poklukar and Janez Žerovnik
On the Double Roman Domination in Generalized Petersen Graphs $P(5k, k)$
Reprinted from: *Mathematics* **2022**, *10*, 119, doi:10.3390/math10010119 87

Wenjie Ning, Yuheng Song and Kun Wang
More on Sombor Index of Graphs
Reprinted from: *Mathematics* **2022**, *10*, 301, doi:10.3390/math10030301 107

Yaser Rowshan, Mostafa Gholami and Stanford Shateyi
A Proof of a Conjecture on Bipartite Ramsey Numbers $B(2, 2, 3)$
Reprinted from: *Mathematics* **2022**, *10*, 701, doi:10.3390/math10050701 119

Yangyang Zhou, Dongyang Zhao, Mingyuan Ma and Jin Xu
Total Coloring of Dumbbell Maximal Planar Graphs
Reprinted from: *Mathematics* **2022**, *10*, 912, doi:10.3390/math10060912 129

Yangyang Zhou, Dongyang Zhao, Mingyuan Ma and Jin Xu
Domination Coloring of Graphs
Reprinted from: *Mathematics* **2022**, *10*, 998, doi:10.3390/math10060998 139

Preface to "Advances in Discrete Applied Mathematics and Graph Theory"

Since its origins in the 18th century, graph theory has been a branch of mathematics that is both motivated by and applied to real world problems. Research in discrete mathematics increased in the latter half of the twentieth century mainly due to the development of digital computers. On the other hand, the advances in technology of digital computers enables extensive application of new ideas from discrete mathematics to real-world problems.

The present reprint contains twelve papers published in the Special Issue "Advances in Discrete Applied Mathematics and Graph Theory, 2021" of the MDPI Mathematics journal, which cover a wide range of topics connected to the theory and applications of Graph Theory and Discrete Applied Mathematics. The focus of the majority of papers is on recent advances in graph theory and applications in chemical graph theory. In particular, the topics studied include bipartite and multipartite Ramsey numbers, graph coloring and chromatic numbers, several varieties of domination (Double Roman, Quasi-Total Roman, Total 3-Roman) and two graph indices of interest in chemical graph theory (Sombor index, generalized ABC index), as well as hyperspaces of graphs and local inclusive distance vertex irregular graphs.

Janez Žerovnik and Darja Rupnik Poklukar
Editors

Article

Total Roman {3}-Domination: The Complexity and Linear-Time Algorithm for Trees

Xinyue Liu, Huiqin Jiang, Pu Wu and Zehui Shao *

Institute of Computing Science and Technology, Guangzhou University, Guangzhou 510006, China; xinyue050420@outlook.com or 2111906061@e.gzhu.edu.cn (X.L.); hq.jiang@hotmail.com or 1111906006@e.gzhu.edu.cn (H.J.); puwu1997@126.com or 2111806056@e.gzhu.edu.cn (P.W.)
* Correspondence: zshao@gzhu.edu.cn

Abstract: For a simple graph $G = (V, E)$ with no isolated vertices, a total Roman {3}-dominating function(TR3DF) on G is a function $f : V(G) \to \{0, 1, 2, 3\}$ having the property that (i) $\sum_{w \in N(v)} f(w) \geq 3$ if $f(v) = 0$; (ii) $\sum_{w \in N(v)} f(w) \geq 2$ if $f(v) = 1$; and (iii) every vertex v with $f(v) \neq 0$ has a neighbor u with $f(u) \neq 0$ for every vertex $v \in V(G)$. The weight of a TR3DF f is the sum $f(V) = \sum_{v \in V(G)} f(v)$ and the minimum weight of a total Roman {3}-dominating function on G is called the total Roman {3}-domination number denoted by $\gamma_{t\{R3\}}(G)$. In this paper, we show that the total Roman {3}-domination problem is NP-complete for planar graphs and chordal bipartite graphs. Finally, we present a linear-time algorithm to compute the value of $\gamma_{t\{R3\}}$ for trees.

Keywords: dominating set; total roman {3}-domination; NP-complete; linear-time algorithm

1. Introduction

Let $G = (V, E)$ be a graph with vertex set $V = V(G)$ and edge set $E = E(G)$. For every vertex $v \in V$, the open neighborhood $N_G(v) = N(v) = \{u \in V(G) : uv \in E(G)\}$ and the closed neighborhood $N_G[v] = N[v] = N(v) \cup \{v\}$. We denote the degree of v by $d_G(v) = d(v) = |N_G(v)|$. A vertex of degree one is called a leaf and its neighbor is a support vertex, and a support vertex is called a strong support if it is adjacent to at least two leaves. Let S_n be a star with order n. A tree T is an acyclic connected graph. $G = (G_1 \cup G_2)$ is a union graph G such that $V(G) = V(G_1) \cup V(G_2)$ and $E(G) = E(G_1) \cup E(G_2)$.

Given a graph G and a positive integer k, assume that $f : V(G) \to \{0, 1, 2, ..., k\}$ is a function, and suppose that $(V_0, V_1, .., V_k)$ is the ordered partition of V introduced by f, where $V_i = \{v \in V(G) : f(v) = i\}$ for $i \in \{0, 1, ..., k\}$. Then we can write $f = (V_0, V_1, .., V_k)$ and $\omega_f(V(G)) = \sum_{v \in V(G)} f(v)$ is the weight of a function f of G.

A subset S of a vertex set $V(G)$ is a dominating set of G if for every vertex $v \in V(G) \setminus S$, there exists a vertex $w \in S$ such that wv is an edge of G. The domination number of G denoted by $\gamma(G)$ is the smallest cardinality of a dominating set S of G [1]. A function $f : V(G) \to \{0, 1\}$ is called a dominating function(DF) on G if every vertex u with $f(u) = 0$ has a vertex $v \in N(u)$ such that $f(v) = 1$ [2]. The dominating set problem(DSP) is to find the domination number of G, which has been deeply and widely studied in recent years [3–7].

A subset S of a vertex set $V(G)$ is a total dominating set of G if $\bigcup_{v \in S} N(v) = V(G)$. The total domination number of G denoted by $\gamma_t(G)$ is the smallest cardinality of a total dominating set S of G [8]. The literature on the subject of total domination in graphs has been surveyed and provided in detail in a recent book [9]. Moreover, Michael A. Henning et al. presented a survey of selected recent results on total domination in graphs [10].

The mathematical concept of Roman domination is originally defined and discussed by Stewart et al. [11] and ReVelle et al. [12]. A Roman dominating function(RDF) on graph G is a function $f : V(G) \to \{0, 1, 2\}$ such that every vertex $v \in V(G)$ for which $f(u) = 0$ is adjacent to at least one vertex u with $f(u) = 2$ [13]. The Roman domination number of

Citation: Liu, X.; Jiang, H.; Wu, P.; Shao, Z. Total Roman {3}-Domination: The Complexity and Linear-Time Algorithm for Trees. *Mathematics* **2021**, *9*, 293. https://doi.org/10.3390/math9030293

Academic Editor: Javier Alcaraz
Received: 22 December 2020
Accepted: 26 January 2021
Published: 2 February 2021

Publisher's Note: MDPI stays neutral with regard to jurisdictional clai-ms in published maps and institutio-nal affiliations.

Copyright: © 2021 by the authors. Licensee MDPI, Basel, Switzerland. This article is an open access article distributed under the terms and conditions of the Creative Commons Attribution (CC BY) license (https://creativecommons.org/licenses/by/4.0/).

G is the minimum weight overall RDFs, denoted by $\gamma_R(G)$ [14]. On the basis of Roman domination, signed Roman domination [15], double Roman domination [16] and total Roman domination [17] have been proposed recently.

The total Roman dominating function(TRDF) on G is an RDF f on G with an additional property that every vertex $v \in V(G)$ with $f(v) \neq 0$ has a neighbor u with $f(u) \neq 0$. Let $\gamma_{tR}(G)$ denote the minimum weight of all TRDFs on G. A TRDF on G with weight $\gamma_{tR}(G)$ is called a $\gamma_{tR}(G)$-function. The conception of TRDF was first defined by Hossein Ahangar et al. [18]. In addition, Nicolás Campanelli et al. studied the total Roman domination number of the lexicographic product of graphs [17] and Chloe Lampman et al. presented some basic results of Edge-Critical Graphs [19].

The Roman {2}-dominating function (also named Italian domination) f [20] introduced by Chellali et al. which is defined as follows: $f : V(G) \to \{0,1,2\}$ has the property that $\sum_{u \in N(v)} f(u) \geq 2$ for $f(v) = 0$ [21]. Chellali et al. proved that the Roman {2}-domination problem is NP-complete for bipartite graphs [21]. Hangdi Chen showed that the Roman {2}-domination problem is NP-complete for split graphs, and gave a linear-time algorithm for finding the minimum weight of Roman {2}-dominating function in block graphs [22]. As a generalization of Roman domination, Michael A. Henning et al. studied the relationship between Roman {2}-domination and dominating set parameters in trees [20].

A Roman {3}-dominating function(R{3}DF) f defined by Mojdeh et al. [23], which is defined as follows: $f : V(G) \to \{0,1,2,3\}$ has the property that for every vertex $v \in V(G)$ with $f(v) \in \{0,1\}$ and $\sum_{u \in N(v)} f(u) \geq 3$. Mojdeh et al. presented an upper bound on the Roman {3}-domination number of a connected graph G, characterized the graphs attaining upper bound and showed that the Roman {3}-domination problem is NP-complete, even restricted to bipartite graphs [23].

The total Roman {3}-domination [24] was studied recently. The total Roman {3}-dominating function(TR3DF) on a graph G is an R{3}DF on G with the additional property that every vertex $v \in V(G)$ with $f(v) \neq 0$ has a neighbor w with $f(w) \neq 0$. The minimum weight of a total Roman {3}-dominating function on G denoted by $\gamma_{t\{R3\}}(G)$ is named the total Roman {3}-domination number of G. A $\gamma_{t\{R3\}}(G)$-function is a total Roman {3}-dominating function on G with weight $\gamma_{t\{R3\}}(G)$. Doost Ali Mojdeh et al. showed the relationship among total Roman {3}-domination, total domination, and total Roman{2}-domination parameters. They also presented an upper bound on the total Roman {3}-domination number of a connected graph G and characterized the graphs arriving this bound. Finally, they investigated that total Roman {3}-domination problem is NP-complete for bipartite graphs [24].

In this paper, we further investigate the complexity of total Roman {3}-domination in planar graphs and chordal bipartite graphs. Moreover, we give a linear-time algorithm to compute the $\gamma_{t\{R3\}}$ for trees which answer the problem that it is possible to construct a polynomial algorithm for computing the number of total Roman {3}-domination for trees [24].

2. Complexity

In this section, we study the complexity of total Roman {3}-domination of graph. We show that the total Roman {3}-domination problem is NP-complete for planar graphs and chordal bipartite graphs. Consider the following decision problem.

Total Roman {3}-Domination Problem TR3DP.
Instance: Graph $G = (V, E)$, and a positive integer m.
Question: Does G have a total Roman {3}-function with weight at most m?

Please note that the dominating set problem is NP-complete for planar graphs [25] and chordal bipartite graphs [26]. We show the NP-completeness results by reducing the well-known NP-complete problem, dominating set, to TR3D.

Let G be a graph on n vertices. Let T_v be the tree with $V(T_v) = \{v, v_a, v_b, v_c, v_d, v_e, v_f, v_p, v_q\}$, $E(T_v) = \{vv_a, v_a v_c, v_c v_e, v_c v_f, vv_b, v_b v_d, v_d v_p, v_d v_q\}$, as depicted in Figure 1.

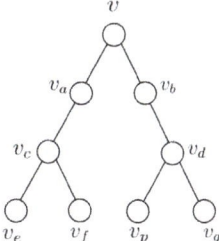

Figure 1. The tree T_v.

Let G' be the graph obtained by adding edges between $v' \in T_{v'}$ and $v'' \in T_{v''}$ if $v'v'' \in E(G)$ from the union of the trees T_v for $v \in V(G)$. Please note that $|V(G')| = n \times |V(T_v)| = 9n$ and $|E(G')| = |E(G)| + n \times |E(T_v)| = |E(G)| + 8n$.

Lemma 1. *If G is a planar graph or chordal bipartite graph, so is G'.*

Lemma 2. *([24]) Let S_n be a star with $n \geq 3$, then $\gamma_{t\{R3\}}(S_n) = 4$.*

Lemma 3. *Let g be a TR3DF of G. If v is a strong support vertex of G, then $\omega_g(N[v]) \geq 4$.*

Proof of Lemma 3. Let $v_1, v_2, .., v_k$ be leaves of v with $k \geq 2$. Since $g(N[v_i]) \geq 3$ for $i \in \{1,2,..,k\}$, we have $g(v_i) \geq 3 - g(v)$ for $i \in \{1,2,..,k\}$. Then $\omega_g(N[v]) = g(v) + \sum_{i \in \{1,2,...,k\}} g(v_i) \geq g(v) + g(v_1) + g(v_2) \geq 6 - g(v)$. If $g(v) \leq 2$, it is clear that $\omega_g(N[v]) \geq 4$. If $g(v) = 3$, there exists a vertex $u \in N(v)$ with $g(u) \neq 0$. Then $\omega_g(N[v]) \geq 4$. □

Lemma 4. *If f is a DF of G with $\omega_f(G) \leq \ell$, then there exists a TR3DF g of G' with $\omega_g(G') \leq \ell + 8n$.*

Proof of Lemma 4. For each $v \in V(G)$, we define g as follows: $V(T_v) \to \{0,1,2,3\}$, $g(v_a) = g(v_b) = 1$, $g(v_c) = g(v_d) = 3$, $g(v) = f(v)$, $g(x) = 0$ otherwise. It is clear that g is a TR3DF of G'. Therefore we have that $\omega_g(G') = \omega_f(G) + 8n \leq \ell + 8n$. □

Claim 1. *Let g be a TR3DF of G', then $\omega_g(T_v') \geq 8$.*

Proof of Claim 1. By Lemmas 2, 3 and definition, we have that $\omega_g(N[v_c]) \geq 4$ and $\omega_g(N[v_d]) \geq 4$. Since $N(v_c) \cap N(v_d) = \emptyset$, then we can reduce $\omega_g(T_v') = \omega_g(N[v_c]) + \omega_g(N[v_d]) \geq 8$. □

Claim 2. *If there exists a TR3DF h of G' with $h(v_a) + h(v_b) \geq 3$ for $v_a, v_b \in V(T_v)$, then there exists a TR3DF g of G' such that $\omega_g(G') \leq \omega_h(G')$ and $g(v_a) + g(v_b) \leq 2$.*

Proof of Claim 2. By the definition of TR3DF, we have $\omega_h(N[v_e]) \geq 3$ and $\omega_h(N[v_p]) \geq 3$, then we have $\omega_h(T_v') \geq 9$.

If $h(v) = 0$, then we define $g : V(G') \to \{0,1,2,3\}$ such that $g(v_e) = g(v_f) = g(v_p) = g(v_q) = 0$, $g(v) = g(v_a) = g(v_b) = 1$, $g(v_c) = g(v_d) = 3$, $g(x) = h(x)$ otherwise, seeing Figure 2. Therefore g is a TR3DF of G' such that $g(v_a) + g(v_b) \leq 2$ and $\omega_g(G') = \omega_h(G')$.

If $h(v) \geq 1$, then we define $g : V(G') \to \{0,1,2,3\}$ such that $g(v_e) = g(v_f) = g(v_p) = g(v_q) = 0$, $g(v_a) = g(v_b) = 1$, $g(v_c) = g(v_d) = 3$, $g(x) = h(x)$ otherwise. Therefore g is a TR3DF of G' such that $g(v_a) + g(v_b) \leq 2$ and $\omega_g(G') \leq \omega_h(G')$. □

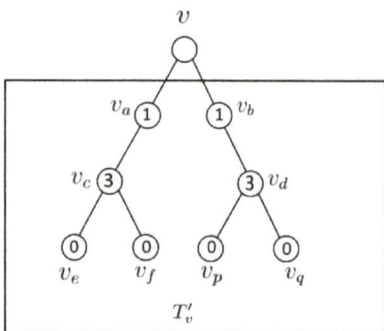

Figure 2. Pre-labeling of g.

Lemma 5. *If g is a TR3DF of G with $\omega_g(G') \leq \ell + 8n$, then there exists a DF f of G with $\omega_f(G) \leq \ell$.*

Proof of Lemma 5. By Claim 2, w.l.o.g, let g be a TR3DF of G' with $g(v_a) + g(v_b) \leq 2$ for $v_a, v_b \in V(T_v)$, $v \in V(G)$. Define $f : V(G) \to \{0,1\}$ such that $f(v) = g(v)$ if $g(v) \leq 1$, and $f(v) = 1$ if $g(v) \geq 2$. For each vertex $v \in V(G)$, since $g(v_a) + g(v_b) \leq 2$, we have $g(v) \geq 1$ or there exists a vertex $u \in N(v) \cap V(G)$ such that $g(u) \geq 1$. Therefore f is DSF of G and $\omega_f(G) \leq \omega_g(G) - 8n \leq \ell$ by Claim 1. □

Theorem 1. *By Lemmas 1, 4, 5, the total Roman {3}-domination problem is NP-complete for planar graphs and chordal bipartite graphs.*

3. A Linear-Time Algorithm for Total Roman {3}-Domination in Trees

In this section, we present a linear-time algorithm to compute the minimum weight of total Roman {3}-dominating function for trees. First, we define the following concepts:

Definition 1. *Let u be a vertex of G, and let $F_{u,G}^{(i,j)}$ on G be a function $f : V(G) \to \{0,1,2,3\}$ having the property that (i) $f(u) = i$, $\sum_{w \in N(u)} f(w) \geq j$; (ii) $\forall v \in V(G) \setminus \{u\}$, $\sum_{p \in N[v]} f(p) \geq 3$ if $f(v) \leq 2$ and $\sum_{p \in N(v)} f(p) \geq 1$ if $f(v) = 3$.*

Definition 2. *The minimum weight overall $F_{u,G}^{(i,j)}$ functions on G denoted by $\gamma_{tR3}^{(i,j)}(u,G)$ is the $F_{u,G}^{(i,j)}$ number of G, and a $\gamma_{tR3}^{(i,j)}(u,G)$-function is an $F_{u,G}^{(i,j)}$ function on G with weight $\gamma_{tR3}^{(i,j)}(u,G)$.*

Definition 3. *Let $coil(x)$ be a function defined as follows:* $coil(x) = \begin{cases} x, x \geq 0; \\ 0, x < 0. \end{cases}$

Lemma 6. *For any graph G with specific vertex u, we have*

$$\gamma_{t\{R3\}}(G) = \min\{\gamma_{tR3}^{(0,3)}(u,G), \gamma_{tR3}^{(1,2)}(u,G), \gamma_{tR3}^{(2,1)}(u,G), \gamma_{tR3}^{(3,1)}(u,G)\}.$$

Lemma 7. *Suppose T_1 and T_2 are trees with specific vertices v and u, respectively. Let T_3 be the tree with the specific vertex u, which is obtained by joining a new edge uv from the union of T_1 and T_2, as depicted in Figure 3.*

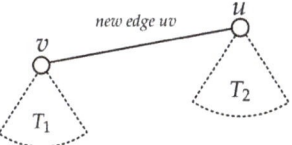

Figure 3. T_3.

Then the following statements hold for $\gamma_{tR3}^{(i,j)}(u, T_k)$.

(a) For $i = 0, j \in \{0, 1, 2, 3\}$, we have:

$$\gamma_{tR3}^{(0,j)}(u, T_3) = min\{\gamma_{tR3}^{(3,1)}(v, T_1) + \gamma_{tR3}^{(0,0)}(u, T_2),$$
$$min\{\gamma_{tR3}^{(s,3-s)}(v, T_1) + \gamma_{tR3}^{(0,coil(j-s))}(u, T_2)|s = 0, 1, 2\}\}$$

(b) For $i \in \{1, 2, 3\}, j \in \{0, 1, 2, 3\}$, we have:

$$\gamma_{tR3}^{(i,j)}(u, T_3) = min\{\gamma_{tR3}^{(s,coil(3-i-s))}(v, T_1) + \gamma_{tR3}^{(i,coil(j-s))}(u, T_2)|s = 0, 1, 2, 3\}$$

Proof of Lemma 7. Let $V(T_1') = V(T_1) \cup \{u\}$, $E(T_1') = E(T_1) \cup \{vu\}$, f be a $\gamma_{tR3}^{(i,j)}(u, G)$-function of T_3, f' be the restriction of f on T_1' and f'' be the restriction of f on T_2.

(a) If f is a $\gamma_{tR3}^{(0,j)}(u, T_3)$-function on T_3, for $j \in \{0, 1, 2, 3\}$. By the definition of $\gamma_{tR3}^{(i,j)}(u, G)$-function, we have that if $f(v) = 3$, then $\sum_{w \in N_{T_3} \setminus \{u\}} f(w) \geq 1$. It follows from the fact that f is a $\gamma_{tR3}^{(0,j)}(u, G)$-function of T_3 if and only if $f = f'' \cup f'$, where at least one of followings holds: (i) f'' is a $\gamma_{tR3}^{(0,0)}(u, G)$-function of T_2, f' is a $\gamma_{tR3}^{(3,1)}(v, T_1)$-function of T_1; (ii) f'' is a $\gamma_{tR3}^{(0,coil(j-s))}(u, G)$-function of T_2, f' is a $\gamma_{tR3}^{(s,3-s)}(v, T_1)$-function of T_1, for $s \in \{0, 1, 2\}$.

(b) It follows from the fact that f is a $\gamma_{tR3}^{(i,j)}(u, T_3)$-function of T_3, for $i \in \{1, 2, 3\}$, $j \in \{0, 1, 2, 3\}$ if and only if $f = f'' \cup f'$, where f'' is a $\gamma_{tR3}^{(i,coil(j-s))}(u, T_2)$-function of T_2 and f' is a $\gamma_{tR3}^{(t,coil(3-i-s))}(v, T_1)$-function of T_1, for $s \in \{0, 1, 2, 3\}$. □

Lemmas 6 and 7 give the following dynamic programming algorithm 1 for the total Roman {3}-domination problem in trees.

Algorithm 1 Counting $\gamma_{t\{R3\}}$ in trees.

Input: A tree T with a tree ordering $[v_1, v_2, .., v_n]$.
Output: the TR3D number $\gamma_{t\{R3\}}(T)$ of T.

1 **for** $p = 1$ to n **do**
2   **for** $i = 0$ to 3, $j = 0$ to 3 **do**
3   **if** $j=0$ **then**
4           $\gamma^{(i,j)}(v_p) \leftarrow i$;
5   **else**
6           $\gamma^{(i,j)}(v_p) \leftarrow \infty$;

7 **for** $p = 1$ to $n - 1$ **do**
8   let v_q be the parent of v_p
9   **for** $i = 0$ to 3 and $j = 0$ to 3 **do**
10   **if** $i=0$ **then**
11          $\gamma^{(i,j)}(v_q) = min\{min\{\gamma^{(s,3-s)}(v_p) + \gamma^{(i,coil(j-s))}(v_q)|s = 0, 1, 2\}; \gamma^{(3,1)}(v_p) + \gamma^{(i,0)}(v_q)\}$;
12   **else**
13          $\gamma^{(i,j)}(v_q) = min\{\gamma^{(s,coil(3-i-s))}(v_p) + \gamma^{(i,coil(j-s))}(v_q)|s = 0, 1, 2, 3\}$;

14 **return** $min\{\gamma^{(0,3)}(v_n), \gamma^{(1,2)}(v_n), \gamma^{(2,1)}(v_n), \gamma^{(3,1)}(v_n)\}$

4. Conclusions

The total Roman {3}-domination problem was introduced and studied in [24], and it was proven to be NP-complete for bipartite graphs. In this paper, we prove that the total Roman {3}-domination problem is NP-complete for planar graphs or chordal bipartite graphs, and showed a linear-time algorithm for total Roman {3}-domination problem on trees. For the algorithmic aspects of the total Roman {3}-domination problem, designing exact algorithms or approximation algorithms on general graphs, or polynomial algorithms for total Roman {3}-domination problem on some special classes graphs deserve further research.

Author Contributions: Conceptualization, X.L., H.J. and Z.S.; writing, X.L. and Z.S.; review, H.J. and Z.S.; investigation: P.W. All authors have contributed equally to this work. All authors have read and agreed to the possible publication of the manuscript.

Funding: This work is supported by the Natural Science Foundation of Guangdong Province under Grant 2018A0303130115.

Conflicts of Interest: The authors declare no conflict of interest.

Abbreviations

The following abbreviations are used in this manuscript:

DF	Dominating function
DSP	Dominating set problem
TRDF	Total Roman dominating function
R3DF	Roman {3}-domination
TR3DF	Total Roman {3}-domination

References

1. Mojdeh, D.A.; Firoozi, P.; Hasni, R. On Connected (γ, k)-critical Graphs. *Australas. J. Comb.* **2010**, *46*, 25–36.
2. Thulasiraman, K.; Swamy, M.N.S. *Graphs: Theory and Algorithms*; Wiley: Hoboken, NJ, USA 1992.
3. Stojmenovic, I.; Seddigh, M.; Zunic, J. Dominating Sets and Neighbor Elimination-Based Broadcasting Algorithms in Wireless Networks. *IEEE Trans. Parallel Distrib. Syst.* **2002**, *13*, 14–25. [CrossRef]
4. Haynes, T.W.; Henning, M.A. *Domination in Graphs*; CRC Press: Boca Raton, FL, USA, 1998.
5. Lund, C.; Yannakakis, M. On the Hardness of Approximating Minimization Problems. *J. ACM* **1994**, *41*, 960–981. [CrossRef]
6. Kinnersley, W.B.; West, D.B.; Zamani, R. Extremal Problems for Game domination Number. *SIAM J. Discret. Math.* **2013**, *27*, 2090–2107. [CrossRef]
7. Haynes, T.W.; Hedetniemi, S.; Slater, P. *Fundamentals of Domination in Graphs*; Marcel Dekker: New York, NY, USA, 1998.
8. Cockayne, E.J.; Dawes, R.M.; Hedetniemi, S.T. Total domination in Graphs. *Networks* **1980**, *10*, 211–219. [CrossRef]
9. Henning, M.A.; Yeo, A. *Total Domination in Graphs*; Springer, Berlin, Germany 2013.
10. Henning, M.A. A Survey of Selected Recent Results On Total domination in Graphs. *Discret. Math.* **2009**, *309*, 32–63. [CrossRef]
11. Stewart, I. Defend the Roman Empire. *Sci. Am.* **1999**, *281*, 136–138. [CrossRef]
12. ReVelle, C.S.; Rosing, K.E. Defendens Imperium Romanum*: A Classical Problem in Military Strategy. *Am. Math. Mon.* **2000**, *107*, 585–594. [CrossRef]
13. Cockayne, E.J.; Dreyer, P.A.; Hedetniemi, S.M.; Hedetniemi, S.T. Roman Domination in Graphs. *Discret. Math.* **2004**, *278*, 11–22. [CrossRef]
14. Chambers, E.W.; Kinnersley, B.; Prince, N.; West, D.B. Extremal Problems for Roman Domination. *SIAM J. Discret. Math.* **2009**, *23*, 1575–1586. [CrossRef]
15. Abdollahzadeh, H.A.; Henning, A.M.; Löwenstein, C.; Zhao, Y.; Samodivkin, V. Signed Roman domination in Graphs. *J. Comb. Optim.* **2014**, *27*, 241–255. [CrossRef]
16. Beeler, R.A.; Haynes, T.W.; Hedetniemi, S.T. Double Roman domination. *Discret. Appl. Math.* **2016**, *211*, 23–29.
17. Campanelli, N.; Kuziak, D. Total Roman domination in the Lexicographic Product of Graphs. *Discret. Appl. Math.* **2019**, *263*, 88–95. [CrossRef]
18. Abdollahzadeh, H.A.; Henning, A.M.; Samodivkin, V.; Yero, G.I. Total Roman domination in Graphs. *Appl. Anal. Discret. Math.* **2016**, *10*, 501–517. [CrossRef]
19. Lampman, C.; Mynhardt, K.; Ogden, S. Total Roman domination Edge-Critical Graphs. *Involv. J. Math.* **2019**, *12*, 1423–1439. [CrossRef]
20. Henning, M.A.; Klostermeyer, W.F. Italian Domination in Trees. *Discret. Appl. Math.* **2017**, *217*, 557–564. [CrossRef]

21. Chellali, M.; Haynes, T.W.; Hedetniemi, S.T.; McRae, A.A. Roman {2}-domination. *Discret. Appl. Math.* **2016**, *204*, 22–28. [CrossRef]
22. Chen, H.; Lu, C. A Note on Roman {2}-domination Problem in Graphs. *arXiv* **2018**, arXiv:1804.09338.
23. Mojdeh, D.A.; Volkmann, L. Roman {3}-domination (Double Italian domination). *Discret. Appl. Math.* **2020**, *283*, 555–564. [CrossRef]
24. Shao, Z.; Mojdeh, D.A.; Volkmann, L. Total Roman {3}-domination in Graphs. *Symmetry* **2020**, *12*, 268. [CrossRef]
25. Zverovich, I.E.; Zverovich, V.E. An Induced Subgraph Characterization of Domination Perfect Graphs. *J. Graph Theory* **1995**, *20*, 375–395. [CrossRef]
26. Müller, H.; Brandstädt, A. The NP-completeness of Steiner Tree and Dominating set for Chordal Bipartite Graphs. *Theor. Comput. Sci.* **1987**, *53*, 257–265. [CrossRef]

Article

The Size, Multipartite Ramsey Numbers for nK_2 Versus Path–Path and Cycle

Yaser Rowshan [1], Mostafa Gholami [1] and Stanford Shateyi [2],*

[1] Department of Mathematics, Institute for Advanced Studies in Basic Sciences (IASBS), Zanjan 66731-45137, Iran; y.rowshan@iasbs.ac.ir (Y.R.); gholami.m@iasbs.ac.ir (M.G.)
[2] Department of Mathematics and Applied Mathematics, School of Mathematical and Natural Sciences, University of Venda, P. Bag X5050, Thohoyandou 0950, South Africa
* Correspondence: stanford.shateyi@univen.ac.za

Abstract: For given graphs $G_1, G_2, \ldots, G_n$ and any integer j, the size of the multipartite Ramsey number $m_j(G_1, G_2, \ldots, G_n)$ is the smallest positive integer t such that any n-coloring of the edges of $K_{j\times t}$ contains a monochromatic copy of G_i in color i for some i, $1 \leq i \leq n$, where $K_{j\times t}$ denotes the complete multipartite graph having j classes with t vertices per each class. In this paper, we computed the size of the multipartite Ramsey numbers $m_j(K_{1,2}, P_4, nK_2)$ for any $j, n \geq 2$ and $m_j(nK_2, C_7)$, for any $j \leq 4$ and $n \geq 2$.

Keywords: Ramsey numbers; multipartite Ramsey numbers; stripes; paths; cycle

MSC: 05D10; 05C55

1. Introduction

In this paper, we were only concerned with undirected, simple and finite graphs. We followed [1] for terminology and notations not defined here. For a given graph G, we denoted its vertex set, edge set, maximum degree and minimum degree by $V(G)$, $E(G)$, $\Delta(G)$ and $\delta(G)$, respectively. For a vertex $v \in V(G)$, we used $\deg_G(v)$ and $N_G(v)$ to denote the degree and neighbours of v in G, respectively. The neighbourhood of a vertex $v \in V(G)$ are denoted by $N_G(v) = \{u \in V(G) \mid uv \in E(G)\}$ and $N_{X_j}(v) = \{u \in V(X_j) \mid uv \in E(G)\}$.

As usual, a cycle and a path on n vertices are denoted by C_n and P_n, respectively. A complete graph on n vertices, denoted K_n, is a graph in which every vertex is adjacent, or connected by an edge, to every other vertex in G. By a stripe mK_2, we mean a graph on $2m$ vertices and m independent edges. A clique is a subset of vertices such that there exists an edge between any pair of vertices in that subset of vertices. An independent set of a graph is a subset of vertices such that there exists no edges between any pair of vertices in that subset. Let C be a set of colors $\{c_1, c_2, \ldots, c_m\}$ and $E(G)$ be the edges of a graph G. An edge coloring $f : E \to C$ assigns each edge in $E(G)$ to a color in C. If an edge coloring uses k color on a graph, then it is known as a k-colored graph. The complete multipartite graph with the partite set $(X_1, X_2, \ldots X_j)$, $|X_i| = s$ for $i = 1, 2, \ldots j$, denoted by $K_{j\times s}$. We use $[X_i, X_j]$ to denote the set of edges between partite sets X_i and X_j. The complement of a graph G, denoted $\overline{G}$, is a graph with the same vertices as G and contains those edges which are not in G. Let $T \subseteq V(G)$ be any subset of vertices of G. Then, the induced subgraph G[T] is the graph whose vertex set is T and whose edge set consists of all of the edges in E(G) that have both endpoints in T.

Since 1956, when Erdös and Rado published the fundamental paper [2], major research has been conducted to compute the size of the multipartite and bipartite Ramsey numbers. A big challenge in combinatorics is to determining the Ramsey numbers for the graphs. We refer to [3] for an overview on Ramsey theory. Ramsey numbers are related to other areas of mathematics, like combinatorial designs [4]. In fact, exact or near-optimal values of several Ramsey numbers depend on the existence of some combinatorial designs like projective planes, which have been studied to date. Many of these connections are briefly described in [3,5]. There are many applications of Ramsey theory in various branches of mathematics and computer science, such as number theory, information theory, set theory, geometry, algebra, topology, logic, ergodic theory and theoretical computer science [6]. In particular, multipartite Ramsey numbers have applications in decision-making problems and communications [7]. There are many mathematicians who present the new results of multipartite Ramsey numbers every year. As a result of this vast range of applications, we were motivated to conduct research on multipartite Ramsey numbers.

For given graphs $G_1, G_2, \ldots, G_n$ and integer j, the size of the multipartite Ramsey number $m_j(G_1, G_2, \ldots, G_n)$ is the smallest integer t such that any n-coloring of the edges of $K_{j \times t}$ contains a monochromatic copy of G_i in color i for some i, $1 \leq i \leq n$, where $K_{j \times t}$ denotes the complete multipartite graph having j classes with t vertices per each class. G is n-colorable to $(G_1, G_2, \ldots, G_n)$ if there exist a t-edge decomposition of G say $(H_1, H_2, \ldots, H_n)$, where $G_i \nsubseteq H_i$ for each $i = 1, 2, \ldots, n$.

The existence of such a positive integer is guaranteed by a result in [2]. The size of the multipartite Ramsey numbers of small paths versus certain classes of graphs have been studied in [8–10]. The size of the multipartite Ramsey numbers of stars versus certain classes of graphs have been studied in [11,12]. In [13,14], Burger, Stipp, Vuuren, and Grobler investigated the multipartite Ramsey numbers $m_j(G_1, G_2)$, where G_1 and G_2 are in a completely balanced multipartite graph, which can be naturally extended to several colors. Recently, the numbers $m_j(G_1, G_2)$ have been investigated for special classes: stripes versus cycles; and stars versus cycles, see [10] and its references. In [15], authors determined the necessary and sufficient conditions for the existence of multipartite Ramsey numbers $m_j(G, H)$ where both G and H are incomplete graphs, which also determined the exact values of the size multipartite Ramsey numbers $m_j(K_{1,m}, K_{1,n})$ for all integers $m, n \geq 1$ and $j = 2, 3$. Syafrizal et al. determined the size multipartite Ramsey numbers of path versus path [16]. $m_3(G, P_3)$ and $m_2(G, P_3)$ where G is a star forest, namely a disjoint union of heterogeneous stars have been studied in [17]. The exact values of the size Ramsey numbers $m_j(P_3, K_{2,n})$ and $m_j(P_4, K_{2,n})$ for $j \geq 3$ computed in [18].

In [12], Lusiani et al. determined the size of the multipartite Ramsey numbers of $m_j(K_{1,m}, H)$, for $j = 2, 3$, where H is a path or a cycle on n vertices, and $K_{1,m}$ is a star of order $m + 1$. In this paper, we computed the size of the multipartite Ramsey numbers $m_j(K_{1,2}, P_4, nK_2)$ for $n, j \geq 2$ and $m_j(nK_2, C_7)$, for $j \leq 4$ and $n \geq 2$ which are the new results of multipartite Ramsey numbers. Computing classic Ramsey numbers is very a difficult problem, therefore we can use multipartite and bipartite Ramsey numbers to obtain an upper bound for a classic Ramsey number. In particular, the first target of this work was to prove the following theorems:

Theorem 1. $m_j(K_{1,2}, P_4, nK_2) = \lfloor \frac{2n}{j} \rfloor + 1$ where $j, n \geq 2$.

In [10], Jayawardene et al. determined the size of the multipartite Ramsey numbers $m_j(nK_2, C_m)$ where $j \geq 2$ and $m \in \{3, 4, 5, 6\}$. The second goal of this work extends these results, as stated below.

Theorem 2. Let $j \in \{2, 3, 4\}$ and $n \geq 2$. Then

$$m_j(nK_2, C_7) = \begin{cases} \infty & j = 2, n \geq 2, \\ 2 & (j, n) = (4, 2), \\ 3 & (j, n) = (3, 2), (4, 3), \\ n & j = 3, n \geq 3, \\ \lceil \frac{n+1}{2} \rceil & j = 4, n \geq 4. \end{cases}$$

We estimate that this value of $m_j(nK_2, C_7)$ holds for every $j \geq 2$. We checked the proof of the main theorems into smaller cases and lemmas in order to simplify the idea of the proof.

2. Proof of Theorem 1

In order to simplify the comprehension, let us split the proof of Theorem 1 into small parts. We begin with a simple but very useful general lower bound in the following lemma:

Lemma 1. $m_j(K_{1,2}, P_4, nK_2) \geq \lfloor \frac{2n}{j} \rfloor + 1$ where $j, n \geq 2$.

Proof. Consider $G = K_{j \times t}$ where $t = \lfloor \frac{2n}{j} \rfloor$ with partition sets X_i, $X_i = \{x_1^i, x_2^i, \ldots, x_t^i\}$ for $i \in \{1, 2, \ldots, j\}$. Consider $x_1^1 \in X_1$, decompose the edges of $K_{j \times t}$ into graphs $G_1, G_2,$ and G_3, where G_1 is a null graph and $G_2 = \overline{G_3}$, where G_3 is $G[X_1 \setminus \{x_1^1\}, X_2, \ldots, X_j]$. In fact G_2 is isomorphic to $K_{1,(j-1)t}$ and:

$$E(G_2) = \{x_1^1 x_i^r \mid r = 2, 3, \ldots, j \text{ and } i = 1, 2 \ldots, t\}.$$

Clearly $E(G_t) \cap E(G_{t'}) = \emptyset$, $E(G) = E(G_1) \cup E(G_2) \cup E(G_3)$, $K_{1,2} \not\subseteq G_1$ and $P_4 \not\subseteq G_2$. Since $|V(K_{j \times t})| = j \times \lfloor \frac{2n}{j} \rfloor \leq 2n$, we have $|V(G_3)| \leq 2n - 1$, that is, $nK_2 \not\subseteq G_3$, which means that $m_3(K_{1,2}, P_4, nK_2) \geq \lfloor \frac{2n}{j} \rfloor + 1$ and the proof is complete. □

Observation 1. Let $G = K_{2,3}$ (or $K_4 - e$). For any subgraph of G, say H, either H has a subgraph isomorphism to $K_{1,2}$ or $\overline{H}$ has a subgraph isomorphism to P_4.

Proof. Let $H \subseteq G = K_{2,3}$, for $G = K_4 - e$ the proof is same. Without loss of generality (w.l.g.), let $X = \{x_1, x_2\}$ and $Y = \{y_1, y_2, y_3\}$ be a partition set of $V(G)$ and P be a maximum path in H. If $|P| \geq 3$, then H has a subgraph isomorphic to $K_{1,2}$, so let $|P| \leq 2$. If $|P| = 1$, then $\overline{H}(= G)$ has a subgraph isomorphic to P_4. Hence, we may assume that $|P| = 2$, w.l.g., and let $P = x_1 y_1$. Since $|P| = 2$, $x_1 y_2, x_1 y_3$ and $x_2 y_1$ are in $E(\overline{H})$ and there is at least one edge of $\{x_2 y_2, x_2 y_3\}$ in $\overline{H}$, in any case, $P_4 \subseteq \overline{H}$ and the proof is complete. □

We determined the exact value of the multipartite Ramsey number of $m_2(K_{1,2}, P_4, nK_2)$ for $n \geq 2$ in the following lemma:

Lemma 2. $m_2(K_{1,2}, P_4, nK_2) = n + 1$ for $n \geq 2$.

Proof. Let $X = \{x_1, x_2, \ldots, x_{n+1}\}$ and $Y = \{y_1, y_2, \ldots, y_{n+1}\}$ be a partition set of $G = K_{n+1,n+1}$. Consider a three-edge coloring G^r, G^b and G^g of G. By Lemma 1, the lower bound holds. Now, let M be the maximum matching in G^g. If $|M| \geq n$, then the lemma holds, so let $|M| \leq n - 1$. If $|M| \leq n - 2$, then we have $K_{3,3} \subseteq \overline{G^g}$ and by Observation 1, the lemma holds, so let $|M| = n - 1$. W.l.g., we may assume that $M = \{x_1 y_1, x_2 y_2, \ldots, x_{n-1} y_{n-1}\}$. By considering the edges between $\{x_n, x_{n+1}\}$ and $Y \setminus \{y_n, y_{n+1}\}$ and the edges between $\{y_n, y_{n+1}\}$ and $X \setminus \{x_n, x_{n+1}\}$, we have $K_{3,2} \subseteq G^r \cup G^b$. Hence, by Observation 1, the lemma holds. □

In the next two lemmas, we consider $m_3(K_{1,2}, P_4, nK_2)$ for certain values of n. In particular, we proved that $m_3(K_{1,2}, P_4, nK_2) = n$, for $n = 2, 3$ in Lemma 3 and $m_3(K_{1,2}, P_4, 4K_2) = 3$ in Lemma 4.

Lemma 3. $m_3(K_{1,2}, P_4, nK_2) = n$ for $n = 2, 3$.

Proof. Let $X_i = \{x_1^i, x_2^i, \ldots, x_n^i\}$ for $i \in \{1, 2, 3\}$ be a partition set of $G = K_{3 \times n}$. Consider a three-edge coloring G^r, G^b and G^g of G. By Lemma 1 the lower bound holds. Now, let M be the maximum matching in G^g and consider the following cases:

Case 1: $n = 2$. If $|M| \geq 2$ then $nK_2 \subseteq G^g$ and the proof is complete. So let $|E(M)| \leq 1$. W.l.g., we may assume that $x_1^1 x_1^2 \in E(M)$, hence, we have $K_4 - e \cong G[x_2^1, x_2^2, X_3] \subseteq G^r \cup G^b$ and by Observation 1, the proof is complete.

Case 2: $n = 3$. In this case, if $|E(M)| \leq 1$ or $|E(M)| \geq 3$, then the proof is the same as case 1. So let $|E(M)| = 2$ and w.l.g., we may assume that $E(M) = \{e_1, e_2\}$—considering any e_1 and e_2 in $E(G)$. In any case, we have $G^r \cup G^b$ has a subgraph isomorphic to $K_{3,2}$, hence, by Observation 1, the lemma holds. Therefore, we have $m_3(K_{1,2}, P_4, 3K_2) = 3$. Now, through cases 1 and 2, the proof is complete. □

Lemma 4. $m_3(K_{1,2}, P_4, 4K_2) = 3$.

Proof. Let $X_i = \{x_1^i, x_2^i, x_3^i\}$ for $i \in \{1, 2, 3\}$ be a partition set of $G = K_{3 \times 3}$. By Lemma 1, the lower bound holds. Consider a three-edge coloring (G^r, G^b, G^g) of G where $4K_2 \not\subseteq G^g$. Let M be a maximum matching in G^g, if $|M| \leq 2$, then the proof is same as Lemma 3. Hence, we may assume that $|M| = 3$ and w.l.g., let $E(M) = \{e_1, e_2, e_3\}$. By Observation 1, there is at least one edge between X_1 and X_2 in G^g, say $e_1 = x_1^1 x_1^2$, and similarly, there is at least one edge between X_3 and $\{x_2^1, x_3^1\}$ in G^g, say $e_2 = x_2^1 x_1^3$, otherwise $K_{3,2} \subseteq G^r \cup G^b$ and the proof is complete. Now, by Observation 1, there is at least one edge between $\{x_3^1, x_2^3, x_3^3\}$ and $\{x_2^2, x_3^2\}$ in G^g, and let e_3 be this edge. If $x_3^1 \notin V(e_3)$ (say $e_3 = x_2^2 x_2^3$), then $K_3 \subseteq G^r \cup G^b[x_3^1, x_3^2, x_3^3]$.

Now, consider the vertex x_1^1 and x_1^2, since $|M| = 3$ and $e_1 = x_1^1 x_1^2$, it is easy to check that $x_1^1 x_3^3, x_1^2 x_3^3 \in E(G^g)$ and $x_1^1 x_3^2, x_1^2 x_3^1 \in E(\overline{G^g})$, otherwise $K_4 - e \subseteq \overline{G^g}$ and the proof is complete. Similarly, we have $x_2^1 x_3^2, x_1^3 x_3^2 \in E(G^g)$ and $x_2^1 x_3^3, x_1^3 x_3^1 \in E(\overline{G^g})$. Now, by considering the edges of $G[X_1, x_1^2, x_2^2, x_1^3, x_3^3]$, it is easy to check that $K_4 - e \subseteq G^r \cup G^b$ and the lemma holds. Hence, we have $x_3^1 \in V(e_3)$ (say $e_3 = x_3^1 x_2^2$), in this case, and we have $K_{2,2} \cong G[x_2^1, x_3^1, x_3^2, x_3^3] \subseteq G^r \cup G^b$, otherwise, if there exists at least one edge between $\{x_2^1, x_3^1\}$ and $\{x_2^2, x_3^2\}$ in G^g, say e, then set $e = e_3$ and the proof is the same. Hence, by considering the vertex x_1^1 and x_1^2, since $|M| = 3$ and $e_1 = x_1^1 x_1^2$, it is easy to check that $K_{3,2} \subseteq G^r \cup G^b$ and by Observation 1 the proof is complete. □

Lemma 5. $m_3(K_{1,2}, P_4, nK_2) \leq \lfloor \frac{2n}{3} \rfloor + 1$ for each $n \geq 2$.

Proof. Let $X_i = \{x_1^i, x_2^i, \ldots, x_t^i\}$ for $i \in \{1, 2, 3\}$ be a partition set of $G = K_{3 \times t}$ where $t = \lfloor \frac{2n}{3} \rfloor + 1$. We will prove this Lemma by induction. For the base step of the induction, since $\lfloor \frac{2 \times 2}{3} \rfloor + 1 = 2$, $\lfloor \frac{2 \times 3}{3} \rfloor + 1 = 3$ and $\lfloor \frac{2 \times 4}{3} \rfloor + 1 = 3$, lemma holds by Lemmas 3 and 4. Suppose that $n \geq 5$ and $m_3(K_{1,2}, P_4, n'K_2) \leq \lfloor \frac{2n'}{3} \rfloor + 1$ for each $n' < n$. We will show that $m_3(K_{1,2}, P_4, nK_2) \leq \lfloor \frac{2n}{3} \rfloor + 1$. By contradiction, we may assume that $m_3(K_{1,2}, P_4, nK_2) > \lfloor \frac{2n}{3} \rfloor + 1$, that is, $K_{3 \times (\lfloor \frac{2n}{3} \rfloor + 1)}$ is three-colorable to $(K_{1,2}, P_4, nK_2)$. Consider a three-edge coloring (G^r, G^b, G^g) of G, such that $K_{1,2} \not\subseteq G^r$, $P_4 \not\subseteq G^b$ and $nK_2 \not\subseteq G^g$. By the induction hypothesis and Lemma 1, we have $m_3(K_{1,2}, P_4, (n-1)K_2) = \lfloor \frac{2(n-1)}{3} \rfloor + 1 \leq \lfloor \frac{2n}{3} \rfloor + 1$. Therefore, since $K_{1,2} \not\subseteq G^r$ and $P_4 \not\subseteq G^b$, we have $(n-1)K_2 \subseteq G^g$. Now, we have the following cases:

Case 1: $\lfloor \frac{2n}{3} \rfloor = \lfloor \frac{2(n-1)}{3} \rfloor + 1$.

Since $\lfloor \frac{2n}{3} \rfloor = \lfloor \frac{2(n-1)}{3} \rfloor + 1$, we have a copy of $H = K_{3 \times (\lfloor \frac{2(n-1)}{3} \rfloor + 1)}$ in G. In other words, for each $i \in \{1, 2, 3\}$, there is a vertex, say $x \in X_i$, such that $x \in V(G) \setminus V(H)$. W.l.g., we may assume that $A = \{x_1^1, x_1^2, x_1^3\}$ would be these vertices. Since $H \subseteq G$, we have $K_{1,2} \not\subseteq G^r[V(H)]$ and $P_4 \not\subseteq G^b[V(H)]$. Hence, by the induction hypothesis, we have $M = (n-1)K_2 \subseteq G^g[V(H)] \subseteq G^g$. We consider that the three vertices do not belong to $V(H)$, i.e., A. Since $nK_2 \not\subseteq G^g$, we have $G[A] \subseteq G^r \cup G^b$. Now, we consider the following Claim:

Claim 1. $n \in B \cup D$ where $B = \{3k \mid k = 1, 2, \ldots\}$ and $D = \{3k + 2 \mid k = 1, 2, \ldots\}$.

Proof of the Claim. By contradiction, we may assume that $n \notin B \cup D$. In other words, let $n = 3k + 1$, then we have:

$$2k = \lfloor \frac{6k}{3} \rfloor = \lfloor \frac{6k}{3} + \frac{2}{3} \rfloor = \lfloor \frac{6k+2}{3} \rfloor = \lfloor \frac{2(3k+1)}{3} \rfloor$$

$$= \lfloor \frac{2n}{3} \rfloor = \lfloor \frac{2(n-1)}{3} \rfloor + 1 = \lfloor \frac{2(3k)}{3} \rfloor + 1 = 2k + 1,$$

which is a contradiction implying that $n \in B \cup D$.

Claim 2. *There is at least one vertex in $V(H) \setminus V(M)$.*

Proof of the Claim. Let $M = (n-1)K_2 \subseteq G^s$, then $|V(M)| = 2(n-1) = 2n - 2$. Since $\lfloor \frac{2n}{3} \rfloor = \lfloor \frac{2(n-1)}{3} \rfloor + 1$, by Claim 1, if $n \in B$, we have $n = 3k$ for $k \geq 2$. Now, we have:

$$\lfloor \frac{2(n-1)}{3} \rfloor + 1 = \lfloor \frac{2(3k-1)}{3} \rfloor + 1 = \lfloor \frac{2(3k)}{3} - \frac{2}{3} \rfloor + 1 = 2k - 1 + 1 = 2k.$$

Hence, we have $|V(H)| = 3 \times (2k) = 6k = 2n$ and thus $|V(H) \setminus V(M)| = 2$. If $n \in D$ then we have:

$$\lfloor \frac{2(n-1)}{3} \rfloor + 1 = \lfloor \frac{2(3k+1)}{3} \rfloor + 1 = \lfloor \frac{2(3k)}{3} + \frac{2}{3} \rfloor + 1 = 2k + 1.$$

Hence, $|V(H)| = 3 \times (2k+1) = 6k + 3 = 2n - 1$. Therefore, $|V(H) \setminus V(M)| = 1$.

By Claim 2, let $x \in V(H) \setminus V(M)$. Since $nK_2 \not\subseteq G^s$, we have $K_4 - e \cong G[A \cup \{x\}] \subseteq G^r \cup G^b$. Hence, by Observation 1, we again have a contradiction.

Case 2: $\lfloor \frac{2n}{3} \rfloor = \lfloor \frac{2(n-1)}{3} \rfloor$.

In this case, by Claim 1 we have $n = 3k + 1$. Since $K_{1,2} \not\subseteq G^r$ and $P_4 \not\subseteq G^b$, by the induction hypothesis, we have $M = (n-1)K_2 \subseteq G^s$. Now, we have the following claim:

Claim 3. $|V(G) \setminus V(M)| = 3$.

Proof of the Claim. Let $M = (n-1)K_2 \subseteq G^s$. Since $|V(X_j)| = \lfloor \frac{2n}{3} \rfloor + 1$ and $n = 3k + 1$, we have $\lfloor \frac{2n}{3} \rfloor + 1 = \lfloor \frac{2(3k+1)}{3} \rfloor + 1 = \lfloor \frac{6k}{3} + \frac{2}{3} \rfloor + 1 = 2k + 1$ and therefore, $|V(G)| = 3 \times (2k+1) = 6k + 3 = 2(3k+1) + 1 = 2n + 1$, that is, $|V(G) \setminus V(M)| = (2n+1) - (2n-2) = 3$.

By Claim 3, we have $|V(G) \setminus V(M)| = 3$. W.l.g., we may assume that $A' = \{x, y, z\}$ has three vertices, since $nK_2 \not\subseteq G^s$, and we have $G[A'] \subseteq G^r \cup G^b$. We consider the three vertices belonging to A', and now, we have the following subcases:

Subcase 2-1: $A' \subseteq X_j$ for only one $j \in \{1, 2, 3\}$. W.l.g. we may assume that $A' \subseteq X_1$ and $E(M) = \{e_i \mid i = 1, 2, \ldots, (n-1)\}$. Since $k \geq 2$ and $3k + 1 = n \geq 7$ we have $|X_j| \geq 5$ and $|E(M) \cap E(G[X_2, X_3])| \geq 3$, otherwise, $K_{3,3} \subseteq G^r \cup G^b$ and by Observation 1; a contradiction. W.l.g. we may assume that $\{x_i^2 x_i^3 \mid i = 1, 2, 3\} \subseteq (E(M) \cap E(G^s[X_2, X_3]))$. Consider $G' = G[A', x_1^2, x_2^2, x_3^2, x_1^3, x_2^3, x_3^3] \cong K_{3 \times 3}$. Since $nK_2 \not\subseteq G^s$, if M' is a maximum matching in G'^s, then $|M'| \leq 3$, otherwise we have $nK_2 = M \setminus \{e_1, e_2, e_3\} \cup M' \subseteq G^s$; a contradiction again. Since $m_3(K_{1,2}, P_4, 4K_2) = 3$ and $|M'| \leq 3$, we have $K_{1,2} \subseteq G'^r \subseteq G^r$ or $P_4 \subseteq G'^b \subseteq G^b$; also a contradiction.

Subcase 2-2: $|A' \cap X_j| = 1$ for each $j \in \{1, 2, 3\}$. W.l.g., we may assume that $x \in X_1, y \in X_2$ and $z \in X_3$. Hence $G[A'] \cong K_3 \subseteq G^r \cup G^b$. Since $|X_j| \geq 5$, we have $|E(M) \cap E(G^s[X_i, X_j])| \geq 2$ for each $i, j \in \{1, 2, 3\}$. W.l.g., we may assume that $x'y' \in E(M) \cap E(G^s[X_1 \setminus \{x\}, X_2 \setminus \{y\}])$, $x' \in X_1$ and $y' \in X_2$. If $x'y$ and $x'z \in E(G^r \cup G^b)$ then we have $K_4 - e \subseteq G^r \cup G^b$ and by Observation 1; a contradiction. So let $x'y$ or $x'z \in E(G^s)$. If $x'y \in E(G^s)$, then, since $nK_2 \not\subseteq G^s$, we have $y'x, y'z \in E(G^r \cup G^b)$, that is, $K_4 - e \subseteq G^r \cup G^b$; we have a contradiction again. So let $x'z \in E(G^s)$ and $x'y \in E(G^r \cup G^b)$. Since $nK_2 \not\subseteq G^s$, we have $y'x \in E(G^r \cup G^b)$. If $|E(G^r) \cap E(G[A'])| \neq 0$, then we have $P_4 \subseteq G^b$. So let $xy, yz, zx \in E(G^b)$ and $xy', yx' \in E(G^r)$. Since $|E(M) \cap E(G^g[X_i, X_j])| \geq 2$ there is at least one edge, say $y''z'' \in E(M) \cap E(G^g[X_2 \setminus \{y\}, X_3 \setminus \{z\}])$. W.l.g., we may assume that $y'' \in X_2$ and $z'' \in X_3$. Since $K_{1,2} \not\subseteq G^r$ and $P_4 \not\subseteq G^b$ we have $y''x, z''y \in E(G^g)$. Hence, we had a $nK_2 = M \setminus \{y''z''\} \cup \{y''x, z''y\}$; a contradiction.

Subcase 2-3: $|A' \cap X_j| = 2$ for only one $j \in \{1, 2, 3\}$. W.l.g., we may assume that $x, y \in X_1$ and $z \in X_2$. Hence, we have $G'[A'] \cong P_3 \subseteq G^r \cup G^b$. Since $k \geq 2$, we have $|X_j| \geq 5$, that is, $|E(M) \cap E(G^g[X_2, X_3])| \geq 3$. W.l.g., we may assume that $vu, v'u' \in E(M) \cap G^g[X_2, X_3]$ where $v, v' \in X_2$ and $u, u' \in X_3$. Now, we have the following claim:

Claim 4. $|N_{G^g}(x) \cap \{v, v'\}| = |N_{G^g}(y) \cap \{v, v'\}| = 0$.

Proof of the Claim. By contradiction, w.l.g., we may assume that $xv \in E(G^g)$. Since $nK_2 \not\subseteq G^g$, we have $yu, zu \in E(G^r \cup G^b)$. Consider $A'' = \{y, z, u\}$ and $M' = M \setminus \{vu\} \cup \{xv\}$. Hence, $M' = (n-1)K_2 \subseteq G^g$ and $|A'' \cap X_j| \neq 0$ for each $j \in \{1, 2, 3\}$; we have a contradiction to subcase 2-2.

Now, by Claim 4, we have $K_{2,3} = G[A' \cup \{v, v'\}] \subseteq G^r \cup G^b$. In this case, by Observation 1, we have $K_{1,2} \subseteq G^r$ or $P_4 \subseteq G^b$; we have a contradiction again.

Therefore, by Cases 1 and 2, we have $m_3(K_{1,2}, P_4, nK_2) \leq \lfloor \frac{2n}{3} \rfloor + 1$ for $n \geq 2$. □

Now, by Lemmas 1 and 5, we have the following lemma:

Lemma 6. $m_3(K_{1,2}, P_4, nK_2) = \lfloor \frac{2n}{3} \rfloor + 1$ for $n \geq 2$.

In the next two lemmas, we consider $m_j(K_{1,2}, P_4, nK_2)$ for each values of $n \geq 2$ and $j \geq 4$. In particular, we proved that $m_j(K_{1,2}, P_4, nK_2) = \lfloor \frac{2n}{j} \rfloor + 1$ for $n \geq 2$ and $j \geq 4$. We started with the following lemma:

Lemma 7. Let $j \geq 4$ and $n \geq 2$. Given that $m_j(K_{1,2}, P_4, (n-1)K_2) = \lfloor \frac{2(n-1)}{j} \rfloor + 1$, it follows that $m_j(K_{1,2}, P_4, nK_2) \leq \lfloor \frac{2n}{j} \rfloor + 1$.

Proof. Let $j \geq 4$ and $n \geq 2$. For $i \in \{1, 2, \ldots, j\}$ let $X_i = \{x_1^i, x_2^i, \ldots, x_t^i\}$ be partition set of $G = K_{j \times t}$ where $t = \lfloor \frac{2n}{j} \rfloor + 1$. Assume that $m_j(K_{1,2}, P_4, (n-1)K_2) = \lfloor \frac{2(n-1)}{j} \rfloor + 1$ is true. To prove $m_j(K_{1,2}, P_4, nK_2) \leq \lfloor \frac{2n}{j} \rfloor + 1$. Consider three-edge coloring (G^r, G^b, G^g) of G. Suppose that $nK_2 \not\subseteq G^g$, we prove that $K_{1,2} \subseteq G^r$ or $P_4 \subseteq G^b$. Let M^* be the maximum matching in G^g. Hence, by the assumption, $|M^*| \leq n - 1$, that is $|V(K_{j \times t}) \cap V(M^*)| \leq 2(n-1)$. Now, we have the following claim:

Claim 5. $|V(K_{j \times t}) \setminus V(M^*)| \geq 3$.

Proof of the Claim. Consider the following cases:
Case 1: Let $2n = jk$ ($2n \equiv 0 \pmod{j}$). In this case, we have:

$$|V(G)| = j \times t = j \times (\lfloor \frac{2n}{j} \rfloor + 1) = j \times \lfloor \frac{2n}{j} \rfloor + j = jk + j = j(k+1).$$

Hence:

$$|V(G) \setminus V(M^*)| \geq j(k+1) - 2(n-1) = jk + j - 2n + 2 = j + 2 \geq 6 \ (j \geq 4).$$

Case 2: Let $2n = jk + r$ ($2n \equiv r \pmod{j}$ where $r \in \{1, 2, \ldots, j-1\}$). In this case, we have:

$$|V(G)| = j \times (\lfloor \frac{2n}{j} \rfloor + 1) = j \times (\lfloor \frac{jk+r}{j} \rfloor + 1) = j \times (\lfloor \frac{jk}{j} + \frac{r}{j} \rfloor + 1) = j \times \lfloor \frac{jk}{j} \rfloor + j = jk + j.$$

Hence we have:
$|V(G) \setminus V(M^*)| \geq j(k+1) - 2(n-1) = jk + j - 2n + 2 = jk + j - jk - r + 2 = j - r + 2 \geq 3$.

By Claim 5, G contains three vertices, say x, y and z in $V(K_{j \times t}) \setminus V(M^*)$. Consider the vertex set $\{x, y, z\}$ and let $\{x, y, z\} \subseteq A = V(G) \setminus V(M^*)$. Now, we have the following cases:

Case 1: Let $x \in X_1, y \in X_2$ and $z \in X_3$, where X_i for $i = 1, 2, 3$ are distinct partition sets of $G = K_{j \times t}$. Note that all vertices of A are adjacent to each other in $\overline{G^g}$. Since $t \geq 2$, we have $|X_i| \geq 2$. Consider the partition X_j for $j \geq 4$. Since $|X_j| \geq 2$, if $|A \cap X_j| \geq 1$ for at least one $j \geq 4$, then we have $K_4 \subseteq G^r \cup G^b$ and the proof is complete by Observation 1. Now, let $|A \cap X_j| = 0$ for each $j \geq 4$. Hence, for $x_1^4 \in X_4$ there exists a vertex, say u such that $x_1^4 u \in E(M^*)$. Consider $N_{G^g}(x_1^4) \cap \{x, y, z\}$. If $|N_{G^g}(x_1^4) \cap \{x, y, z\}| \leq 1$, then we have $K_4 - e \subseteq G^r \cup G^b$ and by Observation 1, the proof is complete. Therefore, let $|N_{G^g}(x_1^4) \cap \{x, y, z\}| \geq 2$. W.l.g., we may assume that $\{x, y\} \subseteq N_{G^g}(x_1^4) \cap \{x, y, z\}$. In this case, we have $|N_{G^g}(u) \cap \{x, y, z\}| = 0$. On the contrary, let $xu \in E(G^g)$ and set $M' = M^* \setminus \{x_1^4 u\} \cup \{x_1^4 y, ux\}$. Clearly M' is a match where $|M'| > |M^*|$, which contradicts the maximality of M^*. Hence, we have $|N_{G^g}(u) \cap \{x, y, z\}| = 0$. Therefore, we have $K_4 - e \subseteq G^r \cup G^b[x, y, z, u]$ and, by Observation 1, the proof is complete.

Case 2: Let $x, y \in X_i$ and $z \in X_{i'}$ where $X_i, X_{i'}$ are distinct partition sets of G. W.l.g., let $i = 1$ and $i' = 2$. Consider the partition $X_j (j \neq 1, 2)$. Since $|X_j| \geq 2$, if $|A \cap X_j| \geq 1$, then we have $K_4 - e \subseteq G^r \cup G^b$ and by Observation 1, the proof is complete. So let $|A \cap X_j| = 0$ for each $j \geq 3$. Now, we have the following claim.

Claim 6. Let $e = v_1 v_2 \in E(M^*)$, and w.l.g. let $|N_{G^g}(v_1) \cap \{x, y, z\}| \geq |N_{G^g}(v_2) \cap \{x, y, z\}|$. If $|N_{G^g}(v_1) \cap \{x, y, z\}| \geq 2$, then $|N_{G^g}(v_2) \cap \{x, y, z\}| = 0$. If $|N_{G^g}(v_1) \cap \{x, y, z\}| = |N_{G^g}(v_2) \cap \{x, y, z\}| = 1$, then v_1, v_2 has the same neighbor in $\{x, y, z\}$.

Proof of the Claim. Let $|N_{G^g}(v_1) \cap \{x, y, z\}| \geq 2$. W.l.g., we may assume that $\{w, w'\} \subseteq N_{G^g}(v_1) \cap \{x, y, z\}$. By contradiction, let $|N_{G^g}(v_2) \cap \{x, y, z\}| \neq 0$, w.l.g., let $w'' \in N_{G^g}(v_2) \cap \{x, y, z\}$. In this case, we set $M' = (M^* \setminus \{v_1 v_2\}) \cup \{v_1 w, v_2 w''\}$. Clearly M' is a match with $|M'| > |M^*|$, which contradicts the maximality of M^*. Thus, let $|N_{G^g}(v_i) \cap \{x, y, z\}| = 1$ for $i = 1, 2$, if v_i has a different neighbor, then the proof is same.

Claim 7. There is at least one edge, say $e = u_i u_j \in E(M^*)$, such that $u_i, u_j \notin X_1, X_2$.

Proof of the Claim. If $|X_j| \geq 3$, then there is at least one edge, say $e = u_i u_j \in E(M^*)$, such that $u_i, u_j \notin X_1, X_2$. Otherwise, we have $K_{3,2} \subseteq G^r \cup G^b[X_j, X_{j'}]$ where $j, j' \geq 3$, hence, by Observation 1; we have a contradiction. So, let $|X_j| = 2$. In this case, if $j \geq 5$, then the proof is same. Now, let $j = 4$. We have $|M^*| \leq 2$, that is, $n \leq 3$. Hence, there is at least one vertex, say $w \in (X_3 \cup X_4) \cap A$; a contradiction to $|A \cap X_j| = 0$.

By Claim 7, there is at least one edge, say $e = u_i u_j \in E(M^*)$, such that $u_i, u_j \notin X_1, X_2$. W.l.g., let $e = u_1 u_2 \in E(M^*)$ such that $u_i \notin X_1, X_2$, also, w.l.g., assume that $|N_{G^g}(u_1) \cap \{x, y, z\}| \geq |N_{G^g}(u_2) \cap \{x, y, z\}|$. If $|N_{G^g}(u_1) \cap \{x, y, z\}| \geq 2$, then by Claim 7, we have $|N_{G^g}(u_2) \cap \{x, y, z\}| = 0$. Hence, we have $K_4 - e \subseteq G^r \cup G^b$. So, let $|N_{G^g}(u_1) \cap \{x, y, z\}| = |N_{G^g}(u_2) \cap \{x, y, z\}| = 1$, in this case, by Claim 7, we have $N_{G^g}(u_1) \cap \{x, y, z\} = N_{G^g}(u_2) \cap \{x, y, z\}$, and if x or y is this vertex, then $K_4 - e \subseteq G^r \cup G^b$; otherwise, $K_{3,2} \subseteq G^r \cup G^b$. In any case, by Observation 1, the proof is complete.

Case 3: Let $x, y, z \in X_i$ where X_i is a partition set of $G = K_{j \times t}$, say $i = 1$. If there exists a vertex, say $w \in X_j \cap A$, where $j \neq 1$, then the proof is the same as Case 2. Hence, let $|A \cap X_j| = 0$. Since $|X_j| \geq 3$, there exists an edge, say $e = vu \in E(M^*)$, such that $v, u \notin X_1$. Consider the neighbors of vertices v and u in X_1. W.l.g., let $|N_{G^g}(v) \cap \{x, y, z\}| \geq |N_{G^g}(u) \cap \{x, y, z\}|$. If $|N_{G^g}(v) \cap \{x, y, z\}| = 0$, then we have $K_{3,2} \subseteq G^r \cup G^b$, so let $|N_{G^g}(v) \cap \{x, y, z\}| \geq 1$. In this case, by Claim 7, we had $|N_{G^g}(u) \cap \{x, y, z\}| \leq 1$. Hence, w.l.g., we may assume that yu and zu be in $E(G^r \cup G^b)$ and $x \in N_{G^g}(v)$. Now, set

$M^{**} = (M^* \setminus \{vu\}) \cup \{vx\}$ and $A' = (A \setminus \{x\}) \cup \{u\}$, the proof is the same as Case 2 and the proof is complete.

According to the Cases 1, 2 and 3 we have $m_j(K_{1,2}, P_4, nK_2) \leq \lfloor \frac{2n}{j} \rfloor + 1$. □

The results of Lemmas 1, 2, 6 and 7, concludes the proof of Theorem 1.

3. Proof of Theorem 2

In this section, we investigate the size multipartite Ramsey numbers $m_j(nK_2, C_7)$ for $j \leq 4$ and $n \geq 2$. In order to simplify the comprehension, let us split the proof of Theorem 2 into small parts. For $j = 2$, since the bipartite graph has no odd cycle, we have $m_2(nK_2, C_7) = \infty$. For other cases, we start with the following proposition:

Proposition 1. $m_3(nK_2, C_7) = 3$ where $n = 2, 3$.

Proof. Clearly, $m_3(nK_2, C_7) \geq 3$. Consider $K_{3 \times 3}$ with the partition set $X_i = \{x_1^i, x_2^i, x_3^i\}$ for $i = 1, 2, 3$. Let G be a subgraph of $K_{3 \times 3}$. For $n = 2$, if $2K_2 \subseteq G$, then proof is complete, so let $2K_2 \nsubseteq G$. In this case, we have $K_{3,2,2} \subseteq \overline{G}$, hence $C_7 \subseteq \overline{G}$, that is, $m_3(2K_2, C_7) = 3$. For $n = 3$ by contradiction, we may assume that $m_3(3K_2, C_7) > 3$, that is, $K_{3 \times 3}$ is 2-colorable to $(3K_2, C_7)$, say $3K_2 \nsubseteq G$ and $C_7 \nsubseteq \overline{G}$. Since $m_3(3K_2, C_6) = 3$ [10], and $3K_2 \nsubseteq G$, we have $C_6 \subseteq \overline{G}$. Let $A = V(C_6)$ and $Y_i = A \cap X_i$ for $i = 1, 2, 3$. If there exists $i \in \{1, 2, 3\}$ such that $|Y_i| = 0$, say $i = 1$, then we have $A = X_2 \cup X_3$ and $C_6 \subseteq \overline{G}[X_2, X_3]$. Let $C_6 = w_1 w_2 \ldots w_6 w_1$. Since $C_7 \nsubseteq \overline{G}$, for each $x_i \in X_1$ in $\overline{G}$, x_i cannot be adjacent to w_i and w_{i+1} for $i = 1, 2, \ldots, 6$. Hence, we have $|N_{\overline{G}}(x_i) \cap V(C_6)| \geq 3$ for each $x_i \in X_1$. One can easily check that in any case, we have $3K_2 \subseteq G$; a contradiction, hence, let $|Y_i| \geq 1$ for each $i = 1, 2, 3$. Set $B = (|Y_1|, |Y_2|, |Y_3|)$. Now, we have the following cases:

Case 1: $B = (3, 2, 1)$. let $A = X_1 \cup \{x_1^2, x_2^2, x_1^3\}$. In this case, we have $C_6 \cong x_1^1 x_1^2 x_2^1 x_2^2 x_3^1 x_1^3 x_1^1$. Consider the vertex set $A' = V(K_{3 \times 3}) \setminus A = \{x_3^2, x_3^2, x_3^3\}$. Since $C_7 \nsubseteq \overline{G}$, we have $|N_{\overline{G}}(x_2^3) \cap \{x_1^1, x_1^2\}| \leq 1$. Hence, $|N_G(x_2^3) \cap \{x_1^1, x_1^2\}| \geq 1$. W.l.g., let $x_2^3 x_1^1 \in E(G)$. By similarity, we have $|N_G(x_3^3) \cap \{x_2^1, x_2^2\}| \geq 1$ and $|N_G(x_3^2) \cap \{x_3^1, x_1^3\}| \geq 1$, see Figure 1. In any case, we have $3K_2 \subseteq G$; a contradiction again.

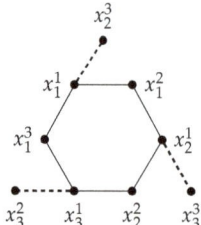

Figure 1. $B = (3, 2, 1)$.

Case 2: $B = (2, 2, 2)$. W.l.g., let $Y_i = \{x_1^i, x_2^i\}$ for $i = 1, 2, 3$. In this case, we have $C_6 \cong w_1 w_2 w_3 w_4 w_5 w_6 w_1$. W.l.g., let $w_1 = x_1^1, w_2 = x_1^2$. Since $|Y_3| = 2$ and $w_4 w_5 \in E(C_6)$, we have $|\{w_3, w_6\} \cap Y_3| \geq 1$. If $|\{w_3, w_6\} \cap Y_3| = 2$, then considering Figure 2a, the proof is the same as case 1. So let $|\{w_3, w_6\} \cap Y_3| = 1$. W.l.g., let $w_3 = x_1^3, x_2^3 = w_5, x_2^1 = w_4, x_2^2 = w_6$. In this case, consider Figure 2b and the proof is the same as case 1. Hence, in any case, we have $3K_2 \subseteq G$; again a contradiction.

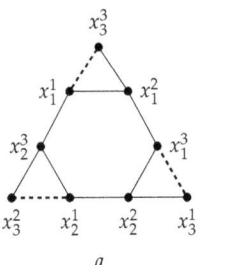
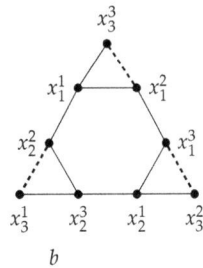

Figure 2. (**a**) $|\{w_3, w_6\} \cap Y_3| = 2$, (**b**) $|\{w_3, w_6\} \cap Y_3| = 1$.

By Cases 1 and 2, we have $3K_2 \subseteq G$. Thus, the proof is complete and the proposition holds. □

We determine the exact value of the multipartite Ramsey number $m_3(nK_2, C_7)$ for $n \geq 3$ in the following lemma:

Lemma 8. *For each $n \geq 3$ we have $m_3(nK_2, C_7) = n$.*

Proof. First, we show that $m_3(nK_2, C_7) \geq n$. Consider the coloring given by $K_{3 \times (n-1)} = G^r \cup G^b$ where $G^r \cong K_{n-1,n-1}$ and $G^b \cong K_{n-1,2(n-1)}$. Since $|V(G^r)| = 2(n-1)$ and G^b is bipartite, we have $nK_2 \nsubseteq G^r$ and $C_7 \nsubseteq G^b$, that is, $m_3(nK_2, C_7) \geq n$. For the upper bound, consider $K_{3 \times n}$ with partite sets $X_i = \{x_1^i, x_2^i, \ldots, x_n^i\}$ for $i = 1, 2, 3$. We will prove this by induction. For $n = 3$, by Proposition 1, the lemma holds. Suppose that $m_3(nK_2, C_7) \leq n$ for each $n \geq 4$. We will show that $m_3((n+1)K_2, C_7) \leq n+1$, as follows: by contradiction, we may assume that $m_3((n+1)K_2, C_7) > n+1$, that is, $K_{3 \times (n+1)}$ is 2-colorable to $((n+1)K_2, C_7)$, say $(n+1)K_2 \nsubseteq G$ and $C_7 \nsubseteq \overline{G}$. Let $X_i' = X_i \setminus \{x_1^i\}$. Hence, by the induction hypothesis, we have $m_3(nK_2, C_7) \leq n$. Therefore, since $|X_i'| = n$ and $C_7 \nsubseteq \overline{G}[X_1', X_2', X_3']$, we have $M = nK_2 \subseteq G[X_1', X_2', X_3']$. If there exists i and j such that $x_1^i x_1^j \in E(G)$, then we have $(n+1)K_2 \subseteq G$; a contradiction. Hence, we have $x_1^i x_1^j \in E(\overline{G})$ for $i, j \in \{1, 2, 3\}$. Let $A = V(K_{3 \times n}) \setminus V(M)$. Hence, we have $|A| = 3n - 2n = n$. Since $(n+1)K_2 \nsubseteq G$, we have $G[A, x_1^1, x_1^2, x_1^3] \subseteq \overline{G}$. Since $|A| = n \geq 4$, one can easily check that, in any case, we have $H \subseteq \overline{G}$, where, $H \in \{K_{5,1,1}, K_{4,2,1}, K_{3,3,1}, K_{3,2,2}\}$. If $H \in \{K_{3,3,1}, K_{3,2,2}\}$, one can easily observe that we have $C_7 \subseteq H \subseteq \overline{G}$; a contradiction again. So let $H \in \{K_{5,1,1}, K_{4,2,1}\}$ and consider the following cases:

Case 1: $A \subseteq X_i$ for only one i, that is, $H = K_{5,1,1}$. W.l.g., let $A \subseteq X_1$ and $\{x_2^1, x_3^1, \ldots, x_5^1\} \subseteq A$. Then, we have $K_{n+1,1,1} \subseteq \overline{G}$ and $M \subseteq G[X_2, X_3]$. Since $n \geq 4$, we have $|M| \geq 4$, that is, there exists at least two edges, say $e_1 = x_1 y_1$ and $e_2 = x_2 y_2$ in $E(M)$, where $\{x_1, x_2, y_1, y_2\} \subseteq X_2 \cup X_3$. W.l.g., let $|N_G(x_i) \cap A| \geq |N_G(y_i) \cap A|$ for $i = 1, 2$. One can easily check that $|N_G(y_i) \cap A| \leq 1$, otherwise, we have $(n+1)K_2 \subseteq G$; a contradiction. Since $|N_G(y_i) \cap A| \leq 1$ and $|A| \geq 5$, we have $|N_{\overline{G}}(y_i) \cap A| \geq 4$. Hence, we have $|N_{\overline{G}}(y_1) \cap N_{\overline{G}}(y_2) \cap A| \geq 3$. W.l.g., we may assume that $\{x_1^1, x_2^1, x_3^1\} \subseteq N_{\overline{G}}(y_1) \cap N_{\overline{G}}(y_2) \cap A$. In this case, we have $C_7 \subseteq \overline{G}[x_1^1, x_2^1, x_3^1, x_1^2, x_1^3, y_1, y_2] \subseteq \overline{G}$; a contradiction again.

Case 2: $H = K_{4,2,1}$. W.l.g., let $|A \cap X_1| = n - 1$ and $|A \cap X_2| = 2$. Let $\{x_2^1, x_3^1, \ldots, x_4^1\} \subseteq A \cap X_1$ and $x_2^2 \in A \cap X_2$, that is, we have $K_{4,2,1} \subseteq K_{n,2,1} = G[A, x_1^1, x_1^2, x_1^3] \subseteq \overline{G}$ and $M \subseteq K_{1,n-1,n}$. That is, there exists at least one edge, say $e = xy$, where $x \in X_2$ and $y \in X_3$. W.l.g., let $|N_G(x) \cap A| \geq |N_G(y) \cap A|$. One can easily check that $|N_G(y) \cap A| \leq 1$. Hence, we have $|N_{\overline{G}}(y) \cap A| \geq 3$ and the proof is same as case 1.

By cases 1 and 2, we have the assumption that $m_3((n+1)K_2, C_7) > n+1$ does not hold. Now we have $m_3(nK_2, C_7) = n$ for each $n \geq 3$. This completes the induction step and the proof. □

Lemma 9. *For $j \geq 3$ and $n \geq j$, we have $m_j(nK_2, C_7) \geq \lceil \frac{2n+2}{j} \rceil$.*

Proof. To show that $m_j(nK_2, C_7) \geq \lceil \frac{2n+2}{j} \rceil$, assume that $\lceil \frac{2n+2}{j} \rceil \geq 1$. Consider the coloring given by $K_{j \times t_0} = G^r \cup G^b$ where $t_0 = \lceil \frac{2n+2}{j} \rceil - 1$ such that $G^r \cong K_{(j-1) \times t_0}$ and $G^b \cong K_{t_0,(j-1)t_0}$. Since G^b is bipartite, we have $C_7 \not\subseteq G^b$, and

$$|V(G^r)| = (j-1) \times t_0 = (j-1)(\lceil \frac{2n+2}{j} \rceil - 1) = (j-1)(\lceil \frac{2n+2}{j} \rceil) - (j-1)$$

$$\leq (j-1)(\frac{2n+2}{j} + 1) - (j-1) = j \times (\frac{2n+2}{j}) - \frac{2n+2}{j}.$$

Since $n \geq j$, we have $|V(G^r)| < 2n$. Hence, we have $nK_2 \not\subseteq G^r$. Since $K_{j \times t_0} = G^r \cup G^b$, we have $m_j(nK_2, C_7) \geq \lceil \frac{2n+2}{j} \rceil$ for $n \geq j \geq 3$. □

Lemma 10. $m_4(4K_2, C_7) = 3$.

Proof. By Lemma 9, we have $m_4(4K_2, C_7) \geq 3$. For the upper bound, consider the coloring given by $K_{4 \times 3} = G^r \cup G^b$ such that $C_7 \not\subseteq G^b$. Since $m_3(3K_2, C_7) = 3$, we have $3K_2 \subseteq G^r[X_1, X_2, X_3] \subseteq G^r$. Let $M = 3K_2$; hence, we have $|V(X_1 \cup X_2 \cup X_3) \setminus V(M)| = 3$. W.l.g., let $A = \{w_1, w_2, w_3\}$ be these vertices. If $E(G^r) \cap E(G[X_4, A]) \neq \emptyset$, then we have $4K_2 \subseteq G^r$. So let $K_{3,3} \subseteq G[X_4, A] \subseteq G^b$. Consider the edge $e = v_1 v_2 \in E(M)$, and it is easy to show that $|N_{G^b}(v_i) \cap X_4| \geq 2$ for some $i \in \{1, 2\}$, otherwise, we have $4K_2 \subseteq G^r$. In any case, one can easily check that $C_7 \subseteq G^b$; which is a contradiction. Thus, we obtain $m_4(4K_2, C_7) = 3$. □

Lemma 11. For $n \geq 4$ we have $m_4(nK_2, C_7) = \lceil \frac{n+1}{2} \rceil$.

Proof. By Lemma 9, we have $m_4(nK_2, C_7) \geq \lceil \frac{n+1}{2} \rceil$. To prove $m_4(nK_2, C_7) \leq \lceil \frac{n+1}{2} \rceil$, consider $K_{4 \times t}$ with partite set $X_i = \{x_1^i, x_2^i, \ldots, x_t^i\}$ for $i = 1, 2, 3, 4$, where $t = \lceil \frac{n+1}{2} \rceil$. We will prove this by induction. For $n = 4$ by Lemma 10, the lemma holds. Now, we consider the following cases:

Case 1: $n = 2k$, where $k \geq 3$. Suppose that $m_4(n'K_2, C_7) \leq \lceil \frac{n'+1}{2} \rceil$ for each $n' < n$. We will show that $m_4(nK_2, C_7) \leq \lceil \frac{n+1}{2} \rceil$ as follows: by contradiction, we may assume that $m_4(nK_2, C_7) > \lceil \frac{n+1}{2} \rceil$, that is, $K_{4 \times t}$ is 2-colorable to (nK_2, C_7), say $nK_2 \not\subseteq G$ and $C_7 \not\subseteq \overline{G}$. Let $X_i' = X_i \setminus \{x_1^i\}$ for $i = 1, 2, 3, 4$. Hence, by the induction hypothesis, we have $m_4((n-1)K_2, C_7) \leq \lceil \frac{n}{2} \rceil = k$. Therefore, since $|X_i'| = k = \frac{n}{2}$ and $C_7 \not\subseteq \overline{G}$, we have $M = (n-1)K_2 \subseteq G[X_1', X_2', X_3', X_4']$. If there exists $i, j \in \{1, 2, 3, 4\}$, where $x_1^i x_1^j \in E(G)$, then $nK_2 \subseteq G$; a contradiction. Now, we have $K_4 \cong \overline{G}[x_1^1, x_1^2, x_1^3, x_1^4] \subseteq \overline{G}^g$. Since $nK_2 \not\subseteq G$ and $\lceil \frac{n+1}{2} \rceil = \lceil \frac{2k+1}{2} \rceil = k + 1$, we have $|V(K_{4 \times k}) \setminus V(M)| = 2n - 2(n-1) = 2$, that is, there exists two vertices, say w_1 and w_2 in $V(K_{4 \times k}) \setminus V(M)$. Since $nK_2 \not\subseteq G$, we have $G[S] \subseteq \overline{G}$, where $S = \{x_1^i \mid i = 1, 2, 3, 4\} \cup \{w_1, w_2\}$. Hence, we have the following claim:

Claim 8. Let $e = v_1 v_2 \in E(M)$ and w.l.g., we may assume that $|N_G(v_1) \cap S| \geq |N_G(v_2) \cap S|$. If $|N_G(v_1) \cap S| \geq 2$ then $|N_G(v_2) \cap S| = 0$. If $|N_G(v_1) \cap S| = 1$ then $|N_G(v_2) \cap S| \leq 1$. If $|N_G(v_i) \cap S| = 1$ then v_1 and v_2 have the same neighbor in S.

Proof of the Claim. By contradiction. We may assume that $\{w, w'\} \subseteq N_G(v_1) \cap S$ and $w'' \in N_G(v_2) \cap S$, in this case, we set $M' = (M \setminus \{v_1 v_2\}) \cup \{v_1 w, v_2 w''\}$. Clearly, M' is a match with $|M'| > |M| = n - 1$, which contradicts the $nK_2 \not\subseteq G$. If $|N_G(v_i) \cap S| = 1$ and v_i has a different neighbor, then the proof is same.

Since $n \geq 4$ and $|M| \geq 3$. If $\{w_1, w_2\} \subseteq X_i$, say X_1, then there is at least one edge, say $e = vu \in E(M)$ such that $v, u \notin X_1$. Otherwise, we have $C_7 \subseteq K_{3 \times 3} \subseteq \overline{G}[X_2, X_3, X_4]$; we again have a contradiction. W.l.g., let $|N_G(v) \cap S| \geq |N_G(u) \cap S|$. Now, by Claim 8 we have $|N_G(u) \cap S| \leq 1$. One can easily check that in any case, we have $C_7 \subseteq \overline{G}[S \cup \{u\}]$; again a contradiction. So w.l.g., let $w_1 \in X_1$ and $w_2 \in X_2$. In this case, since $|N_G(u) \cap S| \leq 1$, we have $C_7 \subseteq \overline{G}[S \cup \{u\}]$; a contradiction again.

Case 2: $n = 2k+1$ where $k \geq 2$, $|X_i| = k+1$. Suppose that $m_4((n-2)K_2, C_7) \leq \lceil \frac{n-2+1}{2} \rceil$ for $n \geq 2$. We show that $m_4(nK_2, C_7) \leq \lceil \frac{n+1}{2} \rceil$ as follows: by contradiction, we may assume that $m_4(nK_2, C_7) > \lceil \frac{n+1}{2} \rceil$, that is, $K_{4 \times t}$ is 2-colorable to (nK_2, C_7), say $nK_2 \not\subseteq G$ and $C_7 \not\subseteq \overline{G}$. Let $X'_i = X_i \setminus \{x_1^i\}$. By the induction hypothesis, we have $m_4((n-2)K_2, C_7) \leq \lceil \frac{n-1}{2} \rceil = \lceil \frac{2k}{2} \rceil = k$. Therefore, since $|X'_i| = k$ and $C_7 \not\subseteq \overline{G}$, we have $M = (n-2)K_2 \subseteq G[X'_1, X'_2, X'_3, X'_4]$ and thus, we have the following claim:

Claim 9. *There exist two edges, say $e_1 = uv$ and $e_2 = u'v'$ in $E(M) = E((n-2)K_2)$, such that v, v', u and u' are in different partites.*

Proof of the Claim. W.l.g., assume that $v \in X'_1$ and $u \in X'_2$. By contradiction, assume that $|E(M) \cap E(G[X'_3, X'_4])| = 0$, that is, $G[X'_3, X'_4] \subseteq \overline{G}$. Since $|V(M)| = 2(n-2)$ and $|X'_i| = k$, we have $|V(M) \cap X'_i| \geq k-2$. Since $k \geq 3$, $|V(M) \cap X'_j| \geq 1$ ($j = 3, 4$). W.l.g., let $e'_j = x_j y_j \in E(M)$ where $x_j \in V(M) \cap X'_j$. W.l.g., we may assume that $y_3 \in V(M) \cap X'_1$. Hence, we have $y_4 \in V(M) \cap X'_1$. In other words, take $e_1 = x_3 y_3$ and $e_2 = x_4 y_4$ and the proof is complete. Hence, we have $|E(M) \cap E(G[X'_2, X'_j])| = 0$ for $j = 3, 4$, in other words, if there exists $e'' \in E(M) \cap E(G[X'_2, X'_j])$, then set $e_1 = e'_1$ and $e_2 = e''$ and the proof is complete. Therefore, for each $e \in E(M)$ we have $v(e) \cap X'_1 \neq \emptyset$ which means that $|M| \leq X'_1 = k$; a contradiction to $|M|$.

Now, by Claim 9 there exist two edges, say $e_1 = uv$ and $e_2 = u'v'$ in $E(M) = E((n-2)K_2)$, such that v, v', u and u' are in different partite. W.l.g., let $e_1 = x_1 x_2$ and $e_2 = x_3 x_4$, since are these edges, and let $x_i \in X'_i$ for $i = 1, 2, 3, 4$. Set $X''_i = X_i \setminus \{x_i\}$, hence, we have $|X''_i| = k$. Since $C_7 \not\subseteq \overline{G}$, we have $C_7 \not\subseteq \overline{G}[X''_1, X''_2, X''_3, X''_4]$. Therefore, by the induction hypothesis, we have $(n-2)K_2 \subseteq G[X''_1, X''_2, X''_3, X''_4]$. Let $M = (n-2)K_2 \subseteq G[X''_1, X''_2, X''_3, X''_4]$, set $M^* = M \cup \{e_1, e_2\}$ hence $|M^*| = n$, that is, $nK_2 \subseteq G$; again a contradiction. Hence, the assumption that $m_4(nK_2, C_7) > \lceil \frac{n+1}{2} \rceil$ does not hold and we have $m_4(nK_2, C_7) \leq \lceil \frac{n+1}{2} \rceil$. This completes the induction step and the proof is complete. By Cases 1 and 2, we have $m_4(nK_2, C_7) = \lceil \frac{n+1}{2} \rceil$ for $n \geq 4$. □

The results of Proposition 1 as well as Lemmas 8 and 11 concludes the proof of Theorem 2.

4. Concluding Remarks and Further Works

There are several papers in which the multipartite Ramsey numbers have been studied. In this paper, as a first target, we compute the size of the multipartite Ramsey number $m_j(K_{1,2}, P_4, nK_2)$ for $n, j \geq 2$. To approach this purpose, we prove four lemmas as follows:

1. $m_j(K_{1,2}, P_4, nK_2) \geq \lfloor \frac{2n}{j} \rfloor + 1$ where $j, n \geq 2$;
2. $m_2(K_{1,2}, P_4, nK_2) = n+1$ for $n \geq 2$;
3. $m_3(K_{1,2}, P_4, nK_2) = \lfloor \frac{2n}{3} \rfloor + 1$ for $n \geq 2$;
4. Let $j \geq 4$ and $n \geq 2$. Given that $m_j(K_{1,2}, P_4, (n-1)K_2) = \lfloor \frac{2(n-1)}{j} \rfloor + 1$, it follows that $m_j(K_{1,2}, P_4, nK_2) \leq \lfloor \frac{2n}{j} \rfloor + 1$.

We computed the size of the multipartite Ramsey numbers $m_j(nK_2, C_7)$, for $j \leq 4$ and $n \geq 2$ as the second purpose of this paper. This extended the result of [10]. To approach this purpose, we proved the following:

1. $m_3(nK_2, C_7) = 3$ where $n = 2, 3$;
2. For each $n \geq 3$ we have $m_3(nK_2, C_7) = n$;
3. For $n \geq 4$ we have $m_4(nK_2, C_7) = \lceil \frac{n+1}{2} \rceil$; We estimated our result for $m_j(nK_2, C_7)$ which holds for every $j \geq 2$, so it could be a good problem to work on.

In addition, one can compute $m_j(K_{1,2}, P_4, m_1 K_2, m_2 K_2)$ and also $m_j(nK_2, C_7)$, for $j \geq 5$ and $n \geq 2$ in the future, using the idea of proofs in this paper.

Author Contributions: These authors contributed equally to this work. All authors have read and agreed to the published version of the manuscript.

Funding: This research received no external funding.

Institutional Review Board Statement: Not applicable.

Informed Consent Statement: Not applicable.

Conflicts of Interest: The authors declare that there is no conflict of interest.

References

1. Bondy, J.A.; Murty, U.S.R. *Graph Theory with Applications*; American Elsevier Publishing Co., Inc.: New York, NY, USA, 1976.
2. Erdös, P.; Rado, R. A partition calculus in set theory. *Bull. Amer. Math. Soc.* **1956**, *62*, 427–489. [CrossRef]
3. Graham, R.L.; Rothschild, B.L.; Spencer, J.H. *Ramsey Theory*; Wiley Series in Discrete Mathematics and Optimization; John Wiley and Sons: Hoboken, NJ, USA, 1990.
4. Erdös, P.; Graham, R.L. On Partition Theorems for Finite Graphs. Infinite and Finite Sets. *Colloq. Math. Soc. János Bolyai N.-Holl. Amsterdam.* **1975**, *10*. Available online: http://citeseerx.ist.psu.edu/viewdoc/summary?doi=10.1.1.210.8857 (accessed on 30 January 2021).
5. Parsons, T. *Ramsey Graph Theory*; Beineke, L.W., Wilson, R.J., Eds.; Selected Topics in Graph Theory; Academic Press: London, UK, 1978; pp. 361–384
6. Rosta, V. Ramsey theory applications. *Electron. J. Comb.* **2004**, *1000*, DS13. [CrossRef]
7. Roberts, F.S. Applications of Ramsey theory. *Discret. Appl. Math.* **1984**, *9*, 251–261. [CrossRef]
8. Łuczak, T.; Polcyn, J. The multipartite Ramsey number for the 3-path of length three. *Discret. Math.* **2018**, *341*, 1270–1274. [CrossRef]
9. Sy, S. On the size multipartite Ramsey numbers for small path versus cocktail party graphs. *Far East J. Appl. Math.* **2011**, *55*, 53–60.
10. Jayawardene, C.; Baskoro, E.T.; Samarasekara, L.; Sy, S. Size multipartite Ramsey numbers for stripes versus small cycles. *Electron. J. Graph Theory Appl. (EJGTA)* **2016**, *4*, 157–170. [CrossRef]
11. Perondi, P.H.; Carmelo, E.L.M. Set and size multipartite Ramsey numbers for stars. *Discrete Appl. Math.* **2018**, *250*, 368–372. [CrossRef]
12. Lusiani, A.; Baskoro, E.T.; Saputro, S.W. On size multipartite Ramsey numbers for stars versus paths and cycles. *Electron. J. Graph Theory Appl. (EJGTA)* **2017**, *5*, 43–50. [CrossRef]
13. Burger, A.P.; Vuuren, J.H.V. Ramsey numbers in complete balanced multipartite graphs. II. Size numbers. *Discret. Math.* **2004**, *283*, 45–49. [CrossRef]
14. Burger, A.P.; Grobler, P.J.P.; Stipp, E.H.; van Vuuren, J.H. Diagonal Ramsey numbers in multipartite graphs. *Util. Math.* **2004**, *66*, 137–163.
15. Lusiani, A.; Baskoro, E.T.; Saputro, S.W. On size multipartite Ramsey numbers for stars. *Indones. J. Comb.* **2020**, *3*, 109–115. [CrossRef]
16. Sy, S.; Baskoro, E.T.; Uttunggadewa, S. The size multipartite Ramsey numbers for paths. *J. Comb. Math. Comb. Comput.* **2005**, *55*, 103.
17. Lusiani, A.; Baskoro, E.T.; Saputro, S.W. On Size Bipartite and Tripartite Ramsey Numbers for The Star Forest and Path on 3 Vertices. *J. Math. Fundam. Sci.* **2020**, *52*, 1–16. [CrossRef]
18. Jayawardene, C. Size Multipartite Ramsey Numbers for Small Paths vs. K2, n. *Ann. Pure Appl. Math.* **2019**, *19*, 7–17. [CrossRef]

Article
Inequalities on the Generalized *ABC* Index

Paul Bosch [1,†], Edil D. Molina [2,†], José M. Rodríguez [3,†] and José M. Sigarreta [2,*,†]

1. Facultad de Ingeniería, Universidad del Desarrollo, Ave. La Plaza 680, San Carlos de Apoquindo, Las Condes, Santiago 7610658, Chile; pbosch@udd.cl
2. Facultad de Matemáticas, Universidad Autónoma de Guerrero, Carlos E. Adame No.54 Col. Garita, 39650 Acalpulco Gro., Mexico; edil941023@gmail.com
3. Departamento de Matemáticas, Universidad Carlos III de Madrid, Avenida de la Universidad 30, 28911 Leganés, Madrid, Spain; jomaro@math.uc3m.es
* Correspondence: josemariasigarretaalmira@hotmail.com or 14366@uagro.mx; Tel.: +52-744-159-2272
† These authors contributed equally to this work.

Abstract: In this work, we obtained new results relating the generalized atom-bond connectivity index with the general Randić index. Some of these inequalities for ABC_α improved, when $\alpha = 1/2$, known results on the ABC index. Moreover, in order to obtain our results, we proved a kind of converse Hölder inequality, which is interesting on its own.

Keywords: ABC index; generalized ABC index; general Randić index; topological indices; converse Hölder inequality

1. Introduction

Mathematical inequalities are at the basis of the processes of approximation, estimation, dimensioning, interpolation, monotonicity, extremes, etc. In general, inequalities appear in models for the study or approach to a certain reality (either objective or subjective). This reason makes it clear that when working with mathematical inequalities, we can essentially find relationships and approximate values of the magnitudes and variables that are associated with one or another practical problem.

In mathematical chemistry, a topological descriptor is a function that associates each molecular graph with a real value; if it correlates well with some chemical property, it is called a topological index. For additional information see [1], for application examples see [2–7].

The atom-bond connectivity index of a graph G was defined in [8] as:

$$ABC(G) = \sum_{uv \in E(G)} \sqrt{\frac{2(d_u + d_v - 2)}{d_u d_v}} = \sqrt{2} \sum_{uv \in E(G)} \sqrt{\frac{d_u + d_v - 2}{d_u d_v}},$$

where uv denotes the edge of the graph G connecting the vertices u and v and d_u is the degree of the vertex u.

The *generalized atom-bond connectivity index* was defined in [9] as:

$$ABC_\alpha(G) = \sum_{uv \in E(G)} \left(\frac{d_u + d_v - 2}{d_u d_v}\right)^\alpha.$$

for any $\alpha \in \mathbb{R} \setminus \{0\}$. Note that $ABC_{1/2} = \frac{\sqrt{2}}{2} ABC$ and ABC_{-3} is the augmented Zagreb index.

There are many papers that have studied the ABC and ABC_α indices (see, e.g., [9–15]). In this paper, we obtained new inequalities relating these indices with the general Randić index. Some of these inequalities for ABC_α improved, when $\alpha = 1/2$, known results on the ABC index. In order to obtain our results, we proved a kind of converse Hölder inequality, Theorem 3, which is interesting on its own.

Citation: Bosch, P.; Molina, E.D.; Rodríguez, J.M.; Sigarreta, J.M. Inequalities on the Generalized *ABC* Index. *Mathematics* **2021**, *9*, 1151. https://doi.org/10.3390/math9101151

Academic Editors: Janez Žerovnik and Darja Rupnik Poklukar

Received: 19 March 2021
Accepted: 17 May 2021
Published: 20 May 2021

Publisher's Note: MDPI stays neutral with regard to jurisdictional claims in published maps and institutional affiliations.

Copyright: © 2021 by the authors. Licensee MDPI, Basel, Switzerland. This article is an open access article distributed under the terms and conditions of the Creative Commons Attribution (CC BY) license (https://creativecommons.org/licenses/by/4.0/).

Throughout this work, a path graph P_n is a tree with n vertices and maximum degree two and a star graph S_n is a tree with n vertices and maximum degree $n-1$.

2. Inequalities Involving ABC_α

In 1998, Bollobás and Erdős [16] generalized the Randić index for $\alpha \in \mathbb{R} \setminus \{0\}$,

$$R_\beta(G) = \sum_{uv \in E(G)} (d_u d_v)^\beta.$$

The general Randić index, also called the *variable Zagreb index* in 2004 by Miličević and Nikolić [17], was extensively studied in [18–20].

The next result relates the ABC_α and R_β indices.

Theorem 1. *Let G be a graph with maximum degree Δ and minimum degree δ and $\alpha > 0$, $\beta \in \mathbb{R} \setminus \{0\}$. Denote by m_2 the cardinality of the set of isolated edges in G.*

(1) If $\beta/\alpha \le -1$ and $\delta > 1$, then:

$$(2\delta - 2)^\alpha \delta^{-2\alpha - 2\beta} R_\beta(G) \le ABC_\alpha(G) \le (2\Delta - 2)^\alpha \Delta^{-2\alpha - 2\beta} R_\beta(G).$$

The equality in each bound is attained if and only if G is a regular graph.

(2) If $\beta/\alpha \le -1$ and $\delta = 1$, then:

$$2^{-\alpha - \beta} (R_\beta(G) - m_2) \le ABC_\alpha(G) \le (2\Delta - 2)^\alpha \Delta^{-2\alpha - 2\beta} (R_\beta(G) - m_2).$$

The equality in the lower bound is attained if and only if G is a union of path graphs P_3 and m_2 isolated edges. The equality in the upper bound is attained if and only if G is a union of a regular graph and m_2 isolated edges.

(3) If $-1 < \beta/\alpha \le -1/2$ and $\delta > 1$, then:

$$(2\delta - 2)^\alpha \delta^{-2\alpha - 2\beta} R_\beta(G) \le ABC_\alpha(G).$$

The equality in the bound is attained if and only if G is a regular graph.

(4) If $-1 < \beta/\alpha \le -1/2$ and $\delta = 1$, then:

$$2^{-\alpha - \beta} (R_\beta(G) - m_2) \le ABC_\alpha(G).$$

The equality in the bound is attained if and only if G is a union of path graphs P_3 and m_2 isolated edges.

(5) If $\beta > 0$ and $\delta > 1$, then:

$$(2\Delta - 2)^\alpha \Delta^{-2\alpha - 2\beta} R_\beta(G) \le ABC_\alpha(G) \le (2\delta - 2)^\alpha \delta^{-2\alpha - 2\beta} R_\beta(G).$$

The equality in each bound is attained if and only if G is a regular graph.

(6) If $\beta > 0$, $\delta = 1$ and $1 + \alpha/\beta \ge \Delta$, then:

$$(2\Delta - 2)^\alpha \Delta^{-2\alpha - 2\beta} (R_\beta(G) - m_2) \le ABC_\alpha(G) \le (\Delta - 1)^\alpha \Delta^{-\alpha - \beta} (R_\beta(G) - m_2).$$

The equality in the lower bound is attained if and only if G is a union of a regular graph and m_2 isolated edges. The equality in the upper bound is attained if and only if G is a union of star graphs $S_{\Delta+1}$ and m_2 isolated edges.

(7) If $\beta > 0$, $\delta = 1$ and $1 + \alpha/\beta \le 2$, then:

$$(2\Delta - 2)^\alpha \Delta^{-2\alpha - 2\beta} (R_\beta(G) - m_2) \le ABC_\alpha(G) \le 2^{-\alpha - \beta} (R_\beta(G) - m_2).$$

The equality in the lower bound is attained if and only if G is a union of a regular graph and m_2 isolated edges. The equality in the upper bound is attained if and only if G is a union of path graphs P_3 and m_2 isolated edges.

(8) If $\beta > 0$, $\delta = 1$ and $2 < 1 + \alpha/\beta < \Delta$, then:

$$(2\Delta - 2)^\alpha \Delta^{-2\alpha - 2\beta}(R_\beta(G) - m_2) \leq ABC_\alpha(G) \leq \frac{\alpha^\alpha \beta^\beta}{(\alpha + \beta)^{\alpha + \beta}}(R_\beta(G) - m_2).$$

The equality in the lower bound is attained if and only if G is a union of a regular graph and m_2 isolated edges. The equality in the upper bound is attained if and only if $\alpha/\beta \in \mathbb{Z}^+$ and G is a union of star graphs $S_{\alpha/\beta+2}$ and m_2 isolated edges.

Proof. First of all, note that $ABC_\alpha(P_2) = 0$ and $R_\beta(P_2) = 1$. Therefore, it suffices to prove the theorem for the case $m_2 = 0$, i.e., when G is a graph without isolated edges. Hence, $\Delta \geq 2$.

We computed the extremal values (for fixed $\lambda \in \mathbb{R}$) of the function $f : [\delta, \Delta] \times ([\delta, \Delta] \setminus [1, 2)) \longrightarrow \mathbb{R}$ given by:

$$f(x, y) = (x + y - 2)(xy)^{-\lambda - 1}.$$

(1) and (2). If $\lambda \leq -1$, then $-\lambda - 1 \geq 0$ and f is a strictly increasing function in each variable, and so,

$$(2\delta - 2)\delta^{-2\lambda - 2} \leq f(x, y) \leq (2\Delta - 2)\Delta^{-2\lambda - 2}.$$

The equality in the lower (respectively, upper) bound is attained if and only if $(x, y) = (\delta, \delta)$ (respectively, $(x, y) = (\Delta, \Delta)$).

If $\delta = 1$, then $f(x, y) \geq f(1, 2) = 2^{-\lambda - 1}$, since $x \in [1, \Delta]$ and $y \in [2, \Delta]$, and the equality in this inequality is attained if and only if $(x, y) = (1, 2)$.

If $\lambda = \beta/\alpha$, then:

$$(2\delta - 2)^\alpha \delta^{-2\beta - 2\alpha}(d_u d_v)^\beta \leq \left(\frac{d_u + d_v - 2}{d_u d_v}\right)^\alpha \leq (2\Delta - 2)^\alpha \Delta^{-2\beta - 2\alpha}(d_u d_v)^\beta$$

for every $uv \in E(G)$ and, consequently,

$$(2\delta - 2)^\alpha \delta^{-2\alpha - 2\beta} R_\beta(G) \leq ABC_\alpha(G) \leq (2\Delta - 2)^\alpha \Delta^{-2\alpha - 2\beta} R_\beta(G).$$

The previous argument shows that the equality in the upper bound is attained if and only if $d_u = d_v = \Delta$ for every $uv \in E(G)$, i.e., G is regular. If $\delta > 1$, then the equality in the lower bound is attained if and only if $d_u = d_v = \delta$ for every $uv \in E(G)$, i.e., G is regular.

If $\lambda = \beta/\alpha$ and $\delta = 1$, then:

$$2^{-\beta - \alpha}(d_u d_v)^\beta \leq \left(\frac{d_u + d_v - 2}{d_u d_v}\right)^\alpha$$

for every $uv \in E(G)$ and, consequently,

$$2^{-\alpha - \beta} R_\beta(G) \leq ABC_\alpha(G).$$

The equality in this bound is attained if and only if $\{d_u, d_v\} = \{1, 2\}$ for every $uv \in E(G)$, i.e., G is a union of path graphs P_3.

(3) and (4). In what follows, by symmetry, we can assume that $x \leq y$. We have:

$$\frac{\partial f}{\partial y}(x, y) = x^{-\lambda - 1}\left(y^{-\lambda - 1} + (x + y - 2)(-\lambda - 1)y^{-\lambda - 2}\right)$$

$$= x^{-\lambda - 1} y^{-\lambda - 2}(y + (x + y - 2)(-\lambda - 1)).$$

If $-1 < \lambda \leq -1/2$, then $-\lambda - 1 \geq -1/2$, and so,

$$\frac{\partial f}{\partial y}(x,y) \geq x^{-\lambda-1}y^{-\lambda-2}\left(y - \frac{x+y-2}{2}\right)$$

$$= x^{-\lambda-1}y^{-\lambda-2}\frac{y-x+2}{2} \geq x^{-\lambda-1}y^{-\lambda-2} > 0.$$

Hence,
$$f(x,y) \geq f(x,x) = (2x-2)x^{-2\lambda-2} = g(x).$$

We have:
$$g'(x) = 2x^{-2\lambda-2} + (2x-2)(-2\lambda-2)x^{-2\lambda-3}$$
$$= 2x^{-2\lambda-3}(x + (x-1)(-2\lambda-2))$$
$$= 2x^{-2\lambda-3}((-2\lambda-1)x + 2\lambda + 2).$$

Since $2\lambda + 2 > 0$ and $-2\lambda - 1 \geq 0$, we have:

$$g'(x) = 2x^{-2\lambda-3}((-2\lambda-1)x + 2\lambda + 2)$$
$$\geq 2x^{-2\lambda-3}(2\lambda + 2) > 0.$$

Thus, $g(x) \geq g(\delta)$ and:
$$f(x,y) \geq g(x) \geq (2\delta-2)\delta^{-2\lambda-2},$$

if $\delta \geq 2$.

If $\lambda = \beta/\alpha$ and $\delta > 1$, then:

$$(2\delta-2)^\alpha \delta^{-2\beta-2\alpha}(d_u d_v)^\beta \leq \left(\frac{d_u + d_v - 2}{d_u d_v}\right)^\alpha$$

for every $uv \in E(G)$ and, consequently,

$$(2\delta-2)^\alpha \delta^{-2\alpha-2\beta} R_\beta(G) \leq ABC_\alpha(G).$$

The previous argument shows that the equality in this bound is attained if and only if $d_u = d_v = \delta$ for every $uv \in E(G)$, i.e., G is regular.

Assume that $\delta = 1$. We proved that $f(x,y) \geq g(x) \geq g(2) = 2^{-2\lambda-1}$ for every $x,y \in [2,\Delta]$. Since $\partial f/\partial y(1,y) > 0$ for every $y \in [2,\Delta]$, we have $f(1,y) \geq f(1,2) = 2^{-\lambda-1}$ for every $y \in [2,\Delta]$. Since $\lambda < 0$, we have $2^{-2\lambda-1} > 2^{-\lambda-1}$ and $f(x,y) \geq 2^{-\lambda-1}$ for every $x \in [1,\Delta] \cap \mathbb{Z}, y \in [2,\Delta] \cap \mathbb{Z}$. Furthermore, the equality in this bound is attained if and only if $(x,y) = (1,2)$.

If $\lambda = \beta/\alpha$, then:

$$2^{-\beta-\alpha}(d_u d_v)^\beta \leq \left(\frac{d_u + d_v - 2}{d_u d_v}\right)^\alpha$$

for every $uv \in E(G)$ and, consequently,

$$2^{-\alpha-\beta} R_\beta(G) \leq ABC_\alpha(G).$$

The equality in this bound is attained if and only if $\{d_u, d_v\} = \{1,2\}$ for every $uv \in E(G)$, i.e., G is a union of path graphs P_3.

(5). Assume now that $\lambda > 0$. Thus, $-\lambda - 1 < -1$ and:

$$\frac{\partial f}{\partial y}(x,y) = x^{-\lambda-1}y^{-\lambda-2}(y + (x+y-2)(-\lambda-1))$$
$$< x^{-\lambda-1}y^{-\lambda-2}(2-x),$$

and:
$$\frac{\partial f}{\partial x}(x,y) < y^{-\lambda-1}x^{-\lambda-2}(2-y).$$

If $\delta > 1$, then f is a strictly decreasing function in each variable, and so,

$$(2\Delta - 2)\Delta^{-2\lambda-2} \leq f(x,y) \leq (2\delta - 2)\delta^{-2\lambda-2}. \quad (1)$$

The equality in the lower (respectively, upper) bound is attained if and only if $(x,y) = (\Delta, \Delta)$ (respectively, $(x,y) = (\delta, \delta)$).

If $\beta > 0$ and $\lambda = \beta/\alpha$, then:

$$(2\Delta - 2)^\alpha \Delta^{-2\beta-2\alpha}(d_u d_v)^\beta \leq \left(\frac{d_u + d_v - 2}{d_u d_v}\right)^\alpha \leq (2\delta - 2)^\alpha \delta^{-2\beta-2\alpha}(d_u d_v)^\beta$$

for every $uv \in E(G)$ and, consequently,

$$(2\Delta - 2)^\alpha \Delta^{-2\alpha-2\beta} R_\beta(G) \leq ABC_\alpha(G) \leq (2\delta - 2)^\alpha \delta^{-2\alpha-2\beta} R_\beta(G).$$

The equality in the lower bound is attained if and only if $d_u = d_v = \Delta$ for every $uv \in E(G)$, i.e., G is regular. Furthermore, the equality in the upper bound is attained if and only if $d_u = d_v = \delta$ for every $uv \in E(G)$, i.e., G is regular.

(6). Note that:

$$\left(\frac{\Delta^2}{2}\right)^{\lambda+1} > \frac{\Delta^2}{2} \geq 2\Delta - 2 \quad \Rightarrow \quad 2^{-\lambda-1} > (2\Delta - 2)\Delta^{-2\lambda-2}. \quad (2)$$

We also have:

$$\Delta^{\lambda+1} > \Delta \geq 2 \quad \Rightarrow \quad (\Delta - 1)\Delta^{-\lambda-1} > (2\Delta - 2)\Delta^{-2\lambda-2}. \quad (3)$$

Assume that $\delta = 1$. If $2 \leq x, y \leq \Delta$, then $f(x,y) \leq f(2,2) = 2^{-2\lambda-1}$. This inequality and the lower bound in (1) give:

$$(2\Delta - 2)\Delta^{-2\lambda-2} \leq f(x,y) \leq 2^{-2\lambda-1}, \quad (4)$$

for every $2 \leq x, y \leq \Delta$.

Let us consider the function $h(y) = f(1,y) = (y-1)y^{-\lambda-1}$ with $2 \leq y \leq \Delta$. We have:

$$h'(y) = -\lambda y^{-\lambda-1} + (\lambda+1)y^{-\lambda-2} = y^{-\lambda-2}(-\lambda y + \lambda + 1),$$

and so, h strictly increases on $(0, 1 + 1/\lambda)$ and strictly decreases on $(1 + 1/\lambda, \infty)$.

If $1 + 1/\lambda \geq \Delta$, then h strictly increases on $(0, \Delta]$ and:

$$2^{-\lambda-1} = h(2) \leq h(y) \leq h(\Delta) = (\Delta - 1)\Delta^{-\lambda-1},$$

for every $2 \leq y \leq \Delta$. These inequalities and Equation (4) give:

$$\min\left\{2^{-\lambda-1}, (2\Delta - 2)\Delta^{-2\lambda-2}\right\} \leq f(x,y) \leq \max\left\{(\Delta - 1)\Delta^{-\lambda-1}, 2^{-2\lambda-1}\right\}.$$

for every $x \in [1, \Delta] \cap \mathbb{Z}, y \in [2, \Delta] \cap \mathbb{Z}$. Since we have in this case $2^{-\lambda-1} = h(2) \leq h(\Delta) = (\Delta - 1)\Delta^{-\lambda-1}$, we conclude:

$$(\Delta - 1)\Delta^{-\lambda-1} \leq \max\left\{(\Delta - 1)\Delta^{-\lambda-1}, 2^{-2\lambda-1}\right\}$$
$$\leq \max\left\{(\Delta - 1)\Delta^{-\lambda-1}, 2^{-\lambda-1}\right\} = (\Delta - 1)\Delta^{-\lambda-1}.$$

Equation (2) gives:

$$\min\left\{2^{-\lambda-1}, (2\Delta - 2)\Delta^{-2\lambda-2}\right\} = (2\Delta - 2)\Delta^{-2\lambda-2}.$$

Hence,
$$(2\Delta - 2)\Delta^{-2\lambda-2} \le f(x,y) \le (\Delta - 1)\Delta^{-\lambda-1},$$
for every $x \in [1,\Delta] \cap \mathbb{Z}, y \in [2,\Delta] \cap \mathbb{Z}$. The equality in the lower (respectively, upper) bound is attained if and only if $(x,y) = (\Delta,\Delta)$ (respectively, $(x,y) = (1,\Delta)$).

If $\beta > 0$ and $\lambda = \beta/\alpha$, then we obtain:
$$(2\Delta - 2)^\alpha \Delta^{-2\beta-2\alpha}(d_u d_v)^\beta \le \left(\frac{d_u + d_v - 2}{d_u d_v}\right)^\alpha \le (\Delta - 1)^\alpha \Delta^{-\beta-\alpha}(d_u d_v)^\beta,$$
$$(2\Delta - 2)^\alpha \Delta^{-2\alpha-2\beta} R_\beta(G) \le ABC_\alpha(G) \le (\Delta - 1)^\alpha \Delta^{-\alpha-\beta} R_\beta(G).$$

The equality in the lower bound is attained if and only if $d_u = d_v = \Delta$ for every $uv \in E(G)$, i.e., G is regular. The equality in the upper bound is attained if and only if $\{d_u, d_v\} = \{1,\Delta\}$ for every $uv \in E(G)$, i.e., G is a union of star graphs $S_{\Delta+1}$.

(7). If $1 + 1/\lambda \le 2$, then h strictly decreases on $[2,\Delta]$ and:
$$(\Delta - 1)\Delta^{-\lambda-1} = h(\Delta) \le h(y) \le h(2) = 2^{-\lambda-1},$$
for every $2 \le y \le \Delta$. These inequalities and Equation (4) give:
$$\min\{(\Delta - 1)\Delta^{-\lambda-1}, (2\Delta - 2)\Delta^{-2\lambda-2}\} \le f(x,y) \le \max\{2^{-\lambda-1}, 2^{-2\lambda-1}\},$$
for every $x \in [1,\Delta] \cap \mathbb{Z}, y \in [2,\Delta] \cap \mathbb{Z}$. Equation (3) gives:
$$(2\Delta - 2)\Delta^{-2\lambda-2} \le f(x,y) \le 2^{-\lambda-1},$$
for every $x \in [1,\Delta] \cap \mathbb{Z}, y \in [2,\Delta] \cap \mathbb{Z}$. The equality in the lower (respectively, upper) bound is attained if and only if $(x,y) = (\Delta,\Delta)$ (respectively, $(x,y) = (1,2)$).

If $\beta > 0$ and $\lambda = \beta/\alpha$, then we obtain for every $uv \in E(G)$:
$$(2\Delta - 2)^\alpha \Delta^{-2\beta-2\alpha}(d_u d_v)^\beta \le \left(\frac{d_u + d_v - 2}{d_u d_v}\right)^\alpha \le 2^{-\beta-\alpha}(d_u d_v)^\beta,$$
$$(2\Delta - 2)^\alpha \Delta^{-2\alpha-2\beta} R_\beta(G) \le ABC_\alpha(G) \le 2^{-\alpha-\beta} R_\beta(G).$$

The equality in the lower bound is attained if and only if $d_u = d_v = \Delta$ for every $uv \in E(G)$, i.e., G is regular. The equality in the upper bound is attained if and only if $\{d_u, d_v\} = \{1,2\}$ for every $uv \in E(G)$, i.e., G is a union of path graphs P_3.

(8). If $2 < 1 + 1/\lambda < \Delta$, then:
$$h(y) \ge \min\{h(2), h(\Delta)\} = \min\{2^{-\lambda-1}, (\Delta - 1)\Delta^{-\lambda-1}\},$$
for every $2 \le y \le \Delta$. Furthermore,
$$h(y) \le h(1 + 1/\lambda) = \frac{1}{\lambda}\left(\frac{\lambda+1}{\lambda}\right)^{-\lambda-1} = \frac{\lambda^\lambda}{(\lambda+1)^{\lambda+1}},$$
for every $2 \le y \le \Delta$. These facts and (4) give:
$$\min\{2^{-\lambda-1}, (\Delta - 1)\Delta^{-\lambda-1}, (2\Delta - 2)\Delta^{-2\lambda-2}\} \le f(x,y)$$
$$\le \max\left\{\frac{\lambda^\lambda}{(\lambda+1)^{\lambda+1}}, 2^{-2\lambda-1}\right\}$$
for every $x \in [1,\Delta] \cap \mathbb{Z}, y \in [2,\Delta] \cap \mathbb{Z}$.

Equations (2) and (3) give:
$$\min\{2^{-\lambda-1}, (\Delta - 1)\Delta^{-\lambda-1}, (2\Delta - 2)\Delta^{-2\lambda-2}\} = (2\Delta - 2)\Delta^{-2\lambda-2}.$$

Since $h(2) \leq h(1+1/\lambda)$, we obtain:

$$2^{-2\lambda-1} < 2^{-\lambda-1} \leq \frac{\lambda^\lambda}{(\lambda+1)^{\lambda+1}},$$

and so,

$$(2\Delta - 2)\Delta^{-2\lambda-2} \leq f(x,y) \leq \frac{\lambda^\lambda}{(\lambda+1)^{\lambda+1}}$$

for every $x \in [1,\Delta] \cap \mathbb{Z}, y \in [2,\Delta] \cap \mathbb{Z}$. The equality in the lower (respectively, upper) bound is attained if and only if $(x,y) = (\Delta, \Delta)$ (respectively, $(x,y) = (1, 1+1/\lambda)$).

If $\beta > 0$ and $\lambda = \beta/\alpha$, then we obtain:

$$\left(\frac{\lambda^\lambda}{(\lambda+1)^{\lambda+1}} \right)^\alpha = \frac{(\beta/\alpha)^\beta}{(\beta/\alpha+1)^{\beta+\alpha}} = \frac{\alpha^\alpha \beta^\beta}{(\alpha+\beta)^{\alpha+\beta}},$$

and we have for every $uv \in E(G)$:

$$(2\Delta - 2)^\alpha \Delta^{-2\beta-2\alpha} (d_u d_v)^\beta \leq \left(\frac{d_u + d_v - 2}{d_u d_v} \right)^\alpha \leq \frac{\alpha^\alpha \beta^\beta}{(\alpha+\beta)^{\alpha+\beta}} (d_u d_v)^\beta,$$

$$(2\Delta - 2)^\alpha \Delta^{-2\alpha-2\beta} R_\beta(G) \leq ABC_\alpha(G) \leq \frac{\alpha^\alpha \beta^\beta}{(\alpha+\beta)^{\alpha+\beta}} R_\beta(G).$$

The equality in the lower bound is attained if and only if $d_u = d_v = \Delta$ for every $uv \in E(G)$, i.e., G is regular. The equality in the upper bound is attained if and only if $\alpha/\beta \in \mathbb{Z}^+$ and $\{d_u, d_v\} = \{1, 1+\alpha/\beta\}$ for every $uv \in E(G)$, i.e., G is a union of star graphs $S_{\alpha/\beta+2}$. □

Note that $ABC_\alpha(G)$ is not well defined if $\alpha < 0$ and G has an isolated edge. The argument in the proof of Theorem 1 gives directly the following result for $\alpha < 0$.

Theorem 2. *Let G be a graph without isolated edges, with maximum degree Δ and minimum degree δ, and $\alpha < 0, \beta \in \mathbb{R} \setminus \{0\}$.*

(1) If $\beta/\alpha \leq -1$ and $\delta > 1$, then:

$$(2\Delta - 2)^\alpha \Delta^{-2\alpha-2\beta} R_\beta(G) \leq ABC_\alpha(G) \leq (2\delta - 2)^\alpha \delta^{-2\alpha-2\beta} R_\beta(G).$$

The equality in each bound is attained if and only if G is a regular graph.

(2) If $\beta/\alpha \leq -1$ and $\delta = 1$, then:

$$(2\Delta - 2)^\alpha \Delta^{-2\alpha-2\beta} R_\beta(G) \leq ABC_\alpha(G) \leq 2^{-\alpha-\beta} R_\beta(G).$$

The equality in the lower bound is attained if and only if G is a regular graph. The equality in the upper bound is attained if and only if G is a union of path graphs P_3.

(3) If $-1 < \beta/\alpha \leq -1/2$ and $\delta > 1$, then:

$$ABC_\alpha(G) \leq (2\delta - 2)^\alpha \delta^{-2\alpha-2\beta} R_\beta(G).$$

The equality in the bound is attained if and only if G is a regular graph.

(4) If $-1 < \beta/\alpha \leq -1/2$ and $\delta = 1$, then:

$$ABC_\alpha(G) \leq 2^{-\alpha-\beta} R_\beta(G).$$

The equality in the bound is attained if and only if G is a union of path graphs P_3.

(5) If $\beta < 0$ and $\delta > 1$, then:

$$(2\delta - 2)^\alpha \delta^{-2\alpha-2\beta} R_\beta(G) \leq ABC_\alpha(G) \leq (2\Delta - 2)^\alpha \Delta^{-2\alpha-2\beta} R_\beta(G).$$

The equality in each bound is attained if and only if G is a regular graph.

(6) If $\beta < 0$, $\delta = 1$ and $1 + \alpha/\beta \geq \Delta$, then:

$$(\Delta - 1)^\alpha \Delta^{-\alpha-\beta} R_\beta(G) \leq ABC_\alpha(G) \leq (2\Delta - 2)^\alpha \Delta^{-2\alpha-2\beta} R_\beta(G).$$

The equality in the lower bound is attained if and only if G is a union of star graphs $S_{\Delta+1}$. The equality in the upper bound is attained if and only if G is a regular graph.

(7) If $\beta < 0$, $\delta = 1$ and $1 + \alpha/\beta \leq 2$, then:

$$2^{-\alpha-\beta} R_\beta(G) \leq ABC_\alpha(G) \leq (2\Delta - 2)^\alpha \Delta^{-2\alpha-2\beta} R_\beta(G).$$

The equality in the lower bound is attained if and only if G is a union of path graphs P_3. The equality in the upper bound is attained if and only if G is a regular graph.

(8) If $\beta < 0$, $\delta = 1$ and $2 < 1 + \alpha/\beta < \Delta$, then:

$$\frac{|\alpha|^\alpha |\beta|^\beta}{|\alpha + \beta|^{\alpha+\beta}} R_\beta(G) \leq ABC_\alpha(G) \leq (2\Delta - 2)^\alpha \Delta^{-2\alpha-2\beta} R_\beta(G).$$

The equality in the lower bound is attained if and only if $\alpha/\beta \in \mathbb{Z}^+$ and G is a union of star graphs $S_{\alpha/\beta+2}$. The equality in the upper bound is attained if and only if G is a regular graph.

Note that Theorems 1 and 2 generalize the classical inequalities:

$$2\sqrt{\delta - 1}\, R(G) \leq ABC(G) \leq 2\sqrt{\Delta - 1}\, R(G). \tag{5}$$

Theorem 1 has the following consequence.

Corollary 1. *Let G be a graph with minimum degree δ and m_2 isolated edges.*

(1) If $\delta > 1$, then:

$$2\sqrt{1 - \frac{1}{\delta}}\, R_{-1/4}(G) \leq ABC(G).$$

The equality in the bound is attained if and only if G is a regular graph.

(2) If $\delta = 1$, then

$$2^{1/4}(R_{-1/4}(G) - m_2) \leq ABC(G).$$

The equality in the bound is attained if and only if G is a union of path graphs P_3 and m_2 isolated edges.

Corollary 1 improves the inequality:

$$2\left(1 - \frac{1}{\sqrt{\delta}}\right) R_{-1/4}(G) \leq ABC(G)$$

in ([21], Theorem 2.5).

In [22], Lemma 4, the following result appeared.

Lemma 1. *Let (X, μ) be a measure space and $f, g : X \to \mathbb{R}$ measurable functions. If there exist positive constants ω, Ω with $\omega |g| \leq |f| \leq \Omega |g|$ μ-a.e., then:*

$$\|f\|_2 \|g\|_2 \leq \frac{1}{2}\left(\sqrt{\frac{\Omega}{\omega}} + \sqrt{\frac{\omega}{\Omega}}\right) \|fg\|_1. \tag{6}$$

If these norms are finite, the equality in the bound is attained if and only if $\omega = \Omega$ and $|f| = \omega |g|$ μ-a.e. or $f = g = 0$ μ-a.e.

We need the following converse Hölder inequality, which is interesting on its own. This result generalizes Lemma 1 and improves the inequality in [23] (Theorem 2).

Theorem 3. *Let (X, μ) be a measure space, $f, g : X \to \mathbb{R}$ measurable functions, and $1 < p, q < \infty$ with $1/p + 1/q = 1$. If there exist positive constants a, b with $a|g|^q \leq |f|^p \leq b|g|^q$ μ-a.e., then:*

$$\|f\|_p \|g\|_q \leq K_p(a,b) \|fg\|_1, \tag{7}$$

with:

$$K_p(a,b) = \begin{cases} \dfrac{1}{p}\left(\dfrac{a}{b}\right)^{1/(2q)} + \dfrac{1}{q}\left(\dfrac{b}{a}\right)^{1/(2p)}, & \text{if } 1 < p < 2, \\[2ex] \dfrac{1}{p}\left(\dfrac{b}{a}\right)^{1/(2q)} + \dfrac{1}{q}\left(\dfrac{a}{b}\right)^{1/(2p)}, & \text{if } p \geq 2. \end{cases}$$

If these norms are finite, the equality in the bound is attained if and only if $a = b$ and $|f|^p = a|g|^q$ μ-a.e. or $f = g = 0$ μ-a.e.

Remark 1. *Since:*

$$K_2(a,b) = \frac{1}{2}\left(\frac{b}{a}\right)^{1/4} + \frac{1}{2}\left(\frac{a}{b}\right)^{1/4},$$

Theorem 3 generalizes Lemma 1 (note that $a = \omega^2$ and $b = \Omega^2$).

Proof. If $p = 2$, then Lemma 1 (with $\omega = a^{1/2}$ and $\Omega = b^{1/2}$) gives the result. Assume now $p \neq 2$, and let us define:

$$k_p(a,b) = \max\left\{ \frac{1}{p}\left(\frac{a}{b}\right)^{1/(2q)} + \frac{1}{q}\left(\frac{b}{a}\right)^{1/(2p)}, \frac{1}{p}\left(\frac{b}{a}\right)^{1/(2q)} + \frac{1}{q}\left(\frac{a}{b}\right)^{1/(2p)} \right\}.$$

We will check at the end of the proof that $k_p(a,b) = K_p(a,b)$.

Let us consider $t \in (0,1)$ and define:

$$G_t(x) := tx^{1-t} + (1-t)x^{-t}$$

for $x > 0$. Since:

$$G'_t(x) = t(1-t)x^{-t} - t(1-t)x^{-t-1} = t(1-t)x^{-t-1}(x-1),$$

G_t is strictly decreasing on $(0,1)$ and strictly increasing on $(1,\infty)$. Thus, if $0 < s \leq S$ are two constants and we consider $s \leq x \leq S$, then:

$$G_t(x) \leq \max\{G_t(s), G_t(S)\} =: A,$$

and if $G_t(x) = A$ for some $s \leq x \leq S$, then $x = s$ or $x = S$.

Note that if $G_t(s) \neq G_t(S)$, the following facts hold: if $G_t(s) > G_t(S)$ and $G_t(x) = A = G_t(s)$, then $x = s$; if $G_t(s) < G_t(S)$ and $G_t(x) = A = G_t(S)$, then $x = S$.

If $x_1, x_2 > 0$ and $sx_2 \leq x_1 \leq Sx_2$, then:

$$t\left(\frac{x_1}{x_2}\right)^{1-t} + (1-t)\left(\frac{x_2}{x_1}\right)^t \leq A,$$

$$tx_1 + (1-t)x_2 \leq A x_1^t x_2^{1-t}.$$

By continuity, this last inequality holds for every $x_1, x_2 \geq 0$ with $sx_2 \leq x_1 \leq Sx_2$. If the equality is attained for some $x_1, x_2 \geq 0$ with $sx_2 \leq x_1 \leq Sx_2$, then $x_1 = sx_2$ or $x_1 = Sx_2$ (the cases $x_1 = 0$ and $x_2 = 0$ are direct).

Choose $t = 1/p$ (thus, $1 - t = 1/q$), $x = x_1^t = x_1^{1/p}$ and $y = x_2^{1-t} = x_2^{1/q}$. Thus,

$$\frac{x^p}{p} + \frac{y^q}{q} \leq Axy \qquad (8)$$

for every $x, y \geq 0$ with $sy^q \leq x^p \leq Sy^q$. If the equality is attained for some $x, y \geq 0$ with $sy^q \leq x^p \leq Sy^q$, then $x^p = sy^q$ or $x^p = Sy^q$.

If $\|f\|_p = 0$ or $\|g\|_q = 0$, then $a|g|^q \leq |f|^p \leq b|g|^q$ μ-a.e. gives $\|f\|_p = \|g\|_q = 0$, and the equality in (7) holds. Assume now that $\|f\|_p \neq 0 \neq \|g\|_q$.

Let us define the function:

$$h := (ab)^{1/(2q)}|g|.$$

We have:

$$\sqrt{\frac{a}{b}}\, h^q = a|g|^q, \qquad \sqrt{\frac{b}{a}}\, h^q = b|g|^q, \qquad \sqrt{\frac{a}{b}}\, h^q \leq |f|^p \leq \sqrt{\frac{b}{a}}\, h^q.$$

If $x = |f|$, $y = h$, $s = (a/b)^{1/2}$, and $S = (b/a)^{1/2}$, then $sh^q \leq |f|^p \leq Sh^q$ and (8) gives:

$$\frac{1}{p}|f|^p + \frac{1}{q}h^q \leq A|f|h.$$

If the equality in this inequality is attained at some point, then:

$$|f|^p = \sqrt{\frac{a}{b}}\, h^q \qquad \text{or} \qquad |f|^p = \sqrt{\frac{b}{a}}\, h^q$$

at that point.

Note that:

$$G_{1/p}(x) = \frac{1}{p} x^{1/q} + \frac{1}{q}\left(\frac{1}{x}\right)^{1/p}$$

and so,

$$A = \max\{G_t(s), G_t(S)\} = \max\left\{G_{1/p}((a/b)^{1/2}), G_{1/p}((b/a)^{1/2})\right\} = k_p(a,b).$$

Hence,

$$\frac{1}{p}|f|^p + \frac{1}{q}h^q \leq k_p(a,b)|f|h,$$

$$\frac{1}{p}\|f\|_p^p + \frac{1}{q}\|h\|_q^q \leq k_p(a,b)\|fh\|_1.$$

Recall that these norms are well defined, although they can be infinite.

If these norms are finite and the equality in the last inequality is attained, then:

$$|f|^p = \sqrt{\frac{a}{b}}\, h^q \qquad \text{or} \qquad |f|^p = \sqrt{\frac{b}{a}}\, h^q$$

μ-a.e. Young's inequality states that:

$$xy \leq \frac{x^p}{p} + \frac{y^q}{q}$$

for every $x, y \geq 0$, and the equality holds if and only if $x^p = y^q$. Thus,

$$\|f\|_p \|h\|_q \leq \frac{1}{p}\|f\|_p^p + \frac{1}{q}\|h\|_q^q \leq k_p(a,b)\|fh\|_1.$$

Therefore, by homogeneity, we conclude:
$$\|f\|_p \|g\|_q \leq k_p(a,b) \|fg\|_1.$$

Let us prove now that $k_p(a,b) = K_p(a,b)$. Consider the function $H_t(x) := G_t(x) - G_t(1/x)$ for $t \in (0,1)$ and $x \in (0,1]$. We have:

$$H_t'(x) = G_t'(x) + \frac{1}{x^2} G_t'\left(\frac{1}{x}\right)$$
$$= t(1-t) x^{-t-1}(x-1) + t(1-t) \frac{1}{x^2} x^{t+1}\left(\frac{1}{x} - 1\right)$$
$$= t(1-t) x^{-t-1}(x-1) + t(1-t) x^{t-2}(1-x)$$
$$= t(1-t)(1-x) x^{-t-1}(x^{2t-1} - 1).$$

If $t \in (0, 1/2)$, then $2t - 1 < 0$ and $H_t'(x) > 0$ for every $x \in (0,1)$, and so, $H_t(x) < H_t(1) = 0$ for every $x \in (0,1)$. Hence, $G_t(x) < G_t(1/x)$ for every $x \in (0,1)$. If $p > 2$ and $a < b$, then $G_{1/p}((a/b)^{1/2}) < G_{1/p}((b/a)^{1/2})$, and:

$$k_p(a,b) = \frac{1}{p}\left(\frac{b}{a}\right)^{1/(2q)} + \frac{1}{q}\left(\frac{a}{b}\right)^{1/(2p)}.$$

If $t \in (1/2, 1)$, then $2t - 1 > 0$ and $H_t'(x) < 0$ for every $x \in (0,1)$, and so, $H_t(x) > H_t(1) = 0$ for every $x \in (0,1)$. Hence, $G_t(x) > G_t(1/x)$ for every $x \in (0,1)$. If $1 < p < 2$ and $a < b$, then $G_{1/p}((a/b)^{1/2}) > G_{1/p}((b/a)^{1/2})$, and:

$$k_p(a,b) = \frac{1}{p}\left(\frac{a}{b}\right)^{1/(2q)} + \frac{1}{q}\left(\frac{b}{a}\right)^{1/(2p)}.$$

Therefore, $k_p(a,b) = K_p(a,b)$.

If $a = b$ and $|f|^p = a|g|^q$ μ-a.e. or $f = g = 0$ μ-a.e., then a computation gives that the equality in (7) is attained.

Finally, assume that the equality in (7) is attained. Seeking for a contradiction, assume that $a \neq b$. The previous argument gives that:

$$|f|^p = \sqrt{\frac{a}{b}} h^q \quad \text{or} \quad |f|^p = \sqrt{\frac{b}{a}} h^q$$

μ-a.e. Since we proved $G_{1/p}((a/b)^{1/2}) \neq G_{1/p}((b/a)^{1/2})$ (recall that $p \neq 2$ and $a < b$), we can conclude that:

$$|f|^p = \sqrt{\frac{a}{b}} h^q \ \mu\text{-a.e.} \quad \text{or} \quad |f|^p = \sqrt{\frac{b}{a}} h^q \ \mu\text{-a.e.}$$

Hence,

$$\|f\|_p^p = \sqrt{\frac{a}{b}} \|h\|_q^q \quad \text{or} \quad \|f\|_p^p = \sqrt{\frac{b}{a}} \|h\|_q^q.$$

Since the equality in Young's inequality gives $\|f\|_p^p = \|h\|_q^q$, we obtain $a = b$, a contradiction. Therefore, $a = b$ and $|f|^p = h^q$ μ-a.e. Hence, $|f|^p = a|g|^q$ μ-a.e. □

Theorem 3 has the following consequence.

Corollary 2. If $1 < p, q < \infty$ with $1/p + 1/q = 1$, $x_j, y_j \geq 0$ and $ay_j^q \leq x_j^p \leq by_j^q$ for $1 \leq j \leq k$ and some positive constants a, b, then:

$$\Big(\sum_{j=1}^k x_j^p\Big)^{1/p} \Big(\sum_{j=1}^k y_j^q\Big)^{1/q} \leq K_p(a,b) \sum_{j=1}^k x_j y_j,$$

where $K_p(a,b)$ is the constant in Theorem 3. If $x_j > 0$ for some $1 \leq j \leq k$, then the equality in the bound is attained if and only if $a = b$ and $x_j^p = ay_j^q$ for every $1 \leq j \leq k$.

The *Platt number* is defined (see, e.g., [24]) as:

$$F(G) = \sum_{uv \in E(G)} (d_u + d_v - 2).$$

Theorem 4. Let G be a graph with m_2 isolated edges and $0 < \alpha < 1$.

(1) Then:
$$ABC_\alpha(G) \leq F(G)^\alpha \big(R_{-\alpha/(1-\alpha)}(G) - m_2\big)^{1-\alpha}.$$

The equality in this bound is attained for the union of any regular or biregular graph and m_2 isolated edges; if G is the union of a connected graph and m_2 isolated edges, then the equality in this bound is attained if and only if G is the union of any regular or biregular connected graph and m_2 isolated edges.

(2) If $\delta > 1$, then:

$$ABC_\alpha(G) \geq \frac{(\Delta-1)^{\alpha/2}\Delta^{\alpha^2/(1-\alpha)}(\delta-1)^{(1-\alpha)/2}\delta^\alpha F(G)^\alpha R_{-\alpha/(1-\alpha)}(G)^{1-\alpha}}{\alpha(\Delta-1)^{1/2}\Delta^{\alpha/(1-\alpha)} + (1-\alpha)(\delta-1)^{1/2}\delta^{\alpha/(1-\alpha)}},$$

if $\alpha \in (0, 1/2]$, and:

$$ABC_\alpha(G) \geq \frac{(\delta-1)^{\alpha/2}\delta^{\alpha^2/(1-\alpha)}(\Delta-1)^{(1-\alpha)/2}\Delta^\alpha F(G)^\alpha R_{-\alpha/(1-\alpha)}(G)^{1-\alpha}}{\alpha(\delta-1)^{1/2}\delta^{\alpha/(1-\alpha)} + (1-\alpha)(\Delta-1)^{1/2}\Delta^{\alpha/(1-\alpha)}},$$

if $\alpha \in (1/2, 1)$. The equality in these bounds is attained if and only if G is regular.

(3) If $\delta = 1$, then:

$$ABC_\alpha(G) \geq \frac{2^\alpha(\Delta-1)^{\alpha/2}\Delta^{\alpha^2/(1-\alpha)} F(G)^\alpha \big(R_{-\alpha/(1-\alpha)}(G) - m_2\big)^{1-\alpha}}{\alpha(2\Delta-2)^{1/2}\Delta^{\alpha/(1-\alpha)} + (1-\alpha)2^{\alpha/(2-2\alpha)}},$$

if $\alpha \in (0, 1/2]$, and:

$$ABC_\alpha(G) \geq \frac{2^{\alpha^2/(2-2\alpha)}\Delta^\alpha(2\Delta-2)^{(1-\alpha)/2} F(G)^\alpha \big(R_{-\alpha/(1-\alpha)}(G) - m_2\big)^{1-\alpha}}{\alpha 2^{\alpha/(2-2\alpha)} + (1-\alpha)(2\Delta-2)^{1/2}\Delta^{\alpha/(1-\alpha)}},$$

if $\alpha \in (1/2, 1)$.

Proof. Since $ABC_\alpha(P_2) = 0$ and $R_\beta(P_2) = 1$, it suffices to prove the theorem for the case $m_2 = 0$, i.e., when G is a graph without isolated edges. Hence, $\Delta \geq 2$.

Hölder's inequality gives:

$$ABC_\alpha(G) = \sum_{uv \in E(G)} \left(\frac{d_u + d_v - 2}{d_u d_v}\right)^\alpha$$

$$\leq \left(\sum_{uv \in E(G)} ((d_u + d_v - 2)^\alpha)^{1/\alpha}\right)^\alpha \left(\sum_{uv \in E(G)} \left(\frac{1}{(d_u d_v)^\alpha}\right)^{1/(1-\alpha)}\right)^{1-\alpha}$$

$$= \left(\sum_{uv \in E(G)} (d_u + d_v - 2)\right)^\alpha \left(\sum_{uv \in E(G)} (d_u d_v)^{-\alpha/(1-\alpha)}\right)^{1-\alpha}$$

$$= F(G)^\alpha R_{-\alpha/(1-\alpha)}(G)^{1-\alpha}.$$

If G is a regular or biregular graph with m edges, then:

$$F(G)^\alpha R_{-\alpha/(1-\alpha)}(G)^{1-\alpha} = ((\Delta + \delta - 2)m)^\alpha \left((\Delta\delta)^{-\alpha/(1-\alpha)} m\right)^{1-\alpha}$$

$$= \frac{(\Delta + \delta - 2)^\alpha}{(\Delta\delta)^\alpha} m = ABC_\alpha(G).$$

Assume that G is connected and that the equality in the first inequality is attained. Hölder's inequality gives that there exists a constant c with:

$$d_u + d_v - 2 = c(d_u d_v)^{-\alpha/(1-\alpha)}$$

for every $uv \in E(G)$. Note that the function $H : [1, \infty) \times [1, \infty) \to [0, \infty)$ given by $H(x, y) = (x + y - 2)(xy)^{\alpha/(1-\alpha)}$ is increasing in each variable. If $uv, uw \in E(G)$, then:

$$c = (d_u + d_v - 2)(d_u d_v)^{\alpha/(1-\alpha)} = (d_u + d_w - 2)(d_u d_w)^{\alpha/(1-\alpha)}$$

implies $d_w = d_v$. Thus, for each vertex $u \in V(G)$, every neighbor of u has the same degree. Since G is a connected graph, this holds if and only if G is regular or biregular.

Assume now that $\delta > 1$. If $\alpha \in (0, 1/2]$, then:

$$K_{1/\alpha}\left((2\delta - 2)\delta^{2\alpha/(1-\alpha)}, (2\Delta - 2)\Delta^{2\alpha/(1-\alpha)}\right)$$

$$= \alpha \left(\frac{\Delta - 1}{\delta - 1}\right)^{(1-\alpha)/2} \left(\frac{\Delta}{\delta}\right)^\alpha + (1 - \alpha)\left(\frac{\delta - 1}{\Delta - 1}\right)^{\alpha/2} \left(\frac{\delta}{\Delta}\right)^{\alpha^2/(1-\alpha)}$$

$$= \frac{\alpha(\Delta - 1)^{(1-\alpha)/2}\Delta^\alpha(\Delta - 1)^{\alpha/2}\Delta^{\alpha^2/(1-\alpha)} + (1-\alpha)(\delta - 1)^{\alpha/2}\delta^{\alpha^2/(1-\alpha)}(\delta - 1)^{(1-\alpha)/2}\delta^\alpha}{(\Delta - 1)^{\alpha/2}\Delta^{\alpha^2/(1-\alpha)}(\delta - 1)^{(1-\alpha)/2}\delta^\alpha}$$

$$= \frac{\alpha(\Delta - 1)^{1/2}\Delta^{\alpha/(1-\alpha)} + (1-\alpha)(\delta - 1)^{1/2}\delta^{\alpha/(1-\alpha)}}{(\Delta - 1)^{\alpha/2}\Delta^{\alpha^2/(1-\alpha)}(\delta - 1)^{(1-\alpha)/2}\delta^\alpha}.$$

If $\alpha \in (1/2, 1)$, then a similar computation gives:

$$K_{1/\alpha}\left((2\delta - 2)\delta^{2\alpha/(1-\alpha)}, (2\Delta - 2)\Delta^{2\alpha/(1-\alpha)}\right)$$

$$= \frac{\alpha(\delta - 1)^{1/2}\delta^{\alpha/(1-\alpha)} + (1-\alpha)(\Delta - 1)^{1/2}\Delta^{\alpha/(1-\alpha)}}{(\delta - 1)^{\alpha/2}\delta^{\alpha^2/(1-\alpha)}(\Delta - 1)^{(1-\alpha)/2}\Delta^\alpha}.$$

Since:

$$(2\delta - 2)\delta^{2\alpha/(1-\alpha)} \leq (d_u + d_v - 2)(d_u d_v)^{\alpha/(1-\alpha)} = \frac{d_u + d_v - 2}{(d_u d_v)^{-\alpha/(1-\alpha)}}$$

$$\leq (2\Delta - 2)\Delta^{2\alpha/(1-\alpha)},$$

Corollary 2 gives:

$$ABC_\alpha(G) = \sum_{uv \in E(G)} \left(\frac{d_u + d_v - 2}{d_u d_v} \right)^\alpha$$

$$\geq \frac{\left(\sum_{uv \in E(G)} (d_u + d_v - 2) \right)^\alpha \left(\sum_{uv \in E(G)} (d_u d_v)^{-\alpha/(1-\alpha)} \right)^{1-\alpha}}{K_{1/\alpha}\left((2\delta - 2)\delta^{2\alpha/(1-\alpha)}, (2\Delta - 2)\Delta^{2\alpha/(1-\alpha)} \right)}$$

$$= \frac{F(G)^\alpha R_{-\alpha/(1-\alpha)}(G)^{1-\alpha}}{K_{1/\alpha}\left((2\delta - 2)\delta^{2\alpha/(1-\alpha)}, (2\Delta - 2)\Delta^{2\alpha/(1-\alpha)} \right)}.$$

This gives the second and third inequalities.
If the graph is regular, then:

$$\frac{F(G)^\alpha R_{-\alpha/(1-\alpha)}(G)^{1-\alpha}}{K_{1/\alpha}\left((2\delta - 2)\delta^{2\alpha/(1-\alpha)}, (2\Delta - 2)\Delta^{2\alpha/(1-\alpha)} \right)}$$

$$= \frac{\left((2\delta - 2)m \right)^\alpha \left(\delta^{-2\alpha/(1-\alpha)} m \right)^{1-\alpha}}{K_{1/\alpha}\left((2\delta - 2)\delta^{2\alpha/(1-\alpha)}, (2\delta - 2)\delta^{2\alpha/(1-\alpha)} \right)}$$

$$= \frac{(2\delta - 2)^\alpha}{\delta^{2\alpha}} m = ABC_\alpha(G).$$

If we have the equality in the second or third inequality, then Corollary 2 gives $(2\delta - 2)\delta^{2\alpha/(1-\alpha)} = (2\Delta - 2)\Delta^{2\alpha/(1-\alpha)}$. Since the function $h(t) = (2t - 2)t^{2\alpha/(1-\alpha)}$ is strictly increasing on $[1, \infty)$, we conclude that $\delta = \Delta$ and G is regular.

Finally, assume that $\delta = 1$. If $\alpha \in (0, 1/2]$, then:

$$K_{1/\alpha}\left(2^{\alpha/(1-\alpha)}, (2\Delta - 2)\Delta^{2\alpha/(1-\alpha)} \right)$$

$$= \alpha (2\Delta - 2)^{(1-\alpha)/2} \left(\frac{\Delta}{2^{1/2}} \right)^\alpha + (1 - \alpha) \left(\frac{1}{2\Delta - 2} \right)^{\alpha/2} \left(\frac{2^{1/2}}{\Delta} \right)^{\alpha^2/(1-\alpha)}$$

$$= \frac{\alpha (2\Delta - 2)^{(1-\alpha)/2} \Delta^\alpha (2\Delta - 2)^{\alpha/2} \Delta^{\alpha^2/(1-\alpha)} + (1 - \alpha) 2^{\alpha^2/(2-2\alpha)} 2^{\alpha/2}}{(2\Delta - 2)^{\alpha/2} \Delta^{\alpha^2/(1-\alpha)} 2^{\alpha/2}}$$

$$= \frac{\alpha (2\Delta - 2)^{1/2} \Delta^{\alpha/(1-\alpha)} + (1 - \alpha) 2^{\alpha/(2-2\alpha)}}{2^\alpha (\Delta - 1)^{\alpha/2} \Delta^{\alpha^2/(1-\alpha)}}.$$

If $\alpha \in (1/2, 1)$, then a similar computation gives:

$$K_{1/\alpha}\left(2^{\alpha/(1-\alpha)}, (2\Delta - 2)\Delta^{2\alpha/(1-\alpha)} \right)$$

$$= \frac{\alpha 2^{\alpha/(2-2\alpha)} + (1 - \alpha)(2\Delta - 2)^{1/2} \Delta^{\alpha/(1-\alpha)}}{2^{\alpha^2/(2-2\alpha)} \Delta^\alpha (2\Delta - 2)^{(1-\alpha)/2}}.$$

Since:

$$2^{\alpha/(1-\alpha)} \leq (d_u + d_v - 2)(d_u d_v)^{\alpha/(1-\alpha)} = \frac{d_u + d_v - 2}{(d_u d_v)^{-\alpha/(1-\alpha)}}$$

$$\leq (2\Delta - 2)\Delta^{2\alpha/(1-\alpha)},$$

Corollary 2 gives:

$$ABC_\alpha(G) = \sum_{uv \in E(G)} \left(\frac{d_u + d_v - 2}{d_u d_v}\right)^\alpha$$

$$\geq \frac{\left(\sum_{uv \in E(G)} (d_u + d_v - 2)\right)^\alpha \left(\sum_{uv \in E(G)} (d_u d_v)^{-\alpha/(1-\alpha)}\right)^{1-\alpha}}{K_{1/\alpha}\left(2^{\alpha/(1-\alpha)}, (2\Delta - 2)\Delta^{2\alpha/(1-\alpha)}\right)}$$

$$= \frac{F(G)^\alpha R_{-\alpha/(1-\alpha)}(G)^{1-\alpha}}{K_{1/\alpha}\left(2^{\alpha/(1-\alpha)}, (2\Delta - 2)\Delta^{2\alpha/(1-\alpha)}\right)}.$$

This gives the fourth and fifth inequalities. □

Theorem 4 has the following consequence.

Corollary 3. *Let G be a graph with m_2 isolated edges.*

(1) Then:

$$ABC(G) \leq \sqrt{2F(G)(R_{-1}(G) - m_2)}.$$

The equality in this bound is attained for the union of any regular or biregular graph and m_2 isolated edges; if G is the union of a connected graph and m_2 isolated edges, then the equality in this bound is attained if and only if G is the union of any regular or biregular connected graph and m_2 isolated edges.

(2) If $\delta > 1$, then:

$$ABC(G) \geq \frac{2\sqrt{2\Delta\delta}\,(\Delta - 1)^{1/4}(\delta - 1)^{1/4} F(G)^{1/2} R_{-1}(G)^{1/2}}{\Delta\sqrt{\Delta - 1} + \delta\sqrt{\delta - 1}}.$$

The equality in this bound is attained if and only if G is regular.

(3) If $\delta = 1$, then:

$$ABC(G) \geq \frac{2\sqrt{2\Delta}\,(\Delta - 1)^{1/4} F(G)^{1/2} (R_{-1}(G) - m_2)^{1/2}}{\Delta\sqrt{\Delta - 1} + 1}.$$

Theorem 5. *If G is a graph with m edges and m_2 isolated edges and $\alpha \in \mathbb{R}$, then:*

$$ABC_\alpha(G) \leq (m - m_2 - 1)^\alpha (R_{-\alpha}(G) - m_2), \quad \text{if } \alpha > 0,$$
$$ABC_\alpha(G) \geq (m - 1)^\alpha R_{-\alpha}(G), \quad \text{if } \alpha < 0 \text{ and } m_2 = 0.$$

The equality in the first bound is attained if and only if G is the union of a star graph and m_2 isolated edges. The equality in the second bound is attained if and only if G is a star graph.

Proof. Since $ABC_\alpha(P_2) = 0$ and $R_\beta(P_2) = 1$, it suffices to prove the theorem for the case $m_2 = 0$, i.e., when G is a graph without isolated edges.

In any graph, the inequality $d_u + d_v \leq m + 1$ holds for every $uv \in E(G)$. If $\alpha > 0$, then:

$$\frac{\left(\frac{d_u+d_v-2}{d_u d_v}\right)^\alpha}{\left(\frac{1}{d_u d_v}\right)^\alpha} = (d_u + d_v - 2)^\alpha \leq (m-1)^\alpha,$$

$$\left(\frac{d_u + d_v - 2}{d_u d_v}\right)^\alpha \leq (m-1)^\alpha (d_u d_v)^{-\alpha},$$

$$ABC_\alpha(G) \leq (m-1)^\alpha R_{-\alpha}(G).$$

If $\alpha < 0$, then we obtain the converse inequality.

If G is a star graph, then $d_u + d_v = m+1$ for every $uv \in E(G)$, and the equality is attained for every α.

If the equality is attained in some inequality, then the previous argument gives that $d_u + d_v = m+1$ for every $uv \in E(G)$. In particular, G is a connected graph. If $m=2$, then $\{d_u, d_v\} = \{1,2\}$ for every $uv \in E(G)$, and so, $G = P_3 = S_3$. Assume now $m \geq 3$. Seeking for a contradiction, assume that $\{d_u, d_v\} \neq \{m, 1\}$ for some $uv \in E(G)$. Since $d_u + d_v = m+1$, we have $2 \leq d_u, d_v \leq m-1$, and so, there exist two different vertices $u', v' \in V(G) \setminus \{u,v\}$ with $uu', vv' \in E(G)$. Since vv' is not incident on u and u', we have $d_u + d_{u'} < m+1$, a contradiction. Hence, $\{d_u, d_v\} = \{m, 1\}$ for every $uv \in E(G)$, and so, G is a star graph. □

Corollary 4. *If G is a graph with m edges and m_2 isolated edges, then:*

$$ABC(G) \leq \sqrt{2(m - m_2 - 1)} \left(R(G) - m_2 \right),$$

and the equality is attained if and only if G is the union of a star graph and m_2 isolated edges.

Note that Theorem 5 (and Corollary 4) improves Items (1) and (2) in Theorems 1 and 2 for many graphs (when $m < 2\Delta - 1$).

3. Conclusions

Topological indices have become a useful tool for the study of theoretical and practical problems in different areas of science. An important line of research associated with topological indices is to find optimal bounds and relations between known topological indices, in particular to obtain bounds for the topological indices associated with invariant parameters of a graph (see [1]).

From the theoretical point of view in this research, a new type of Hölder converse inequality was proposed (Theorem 3 and Corollary 2). From the practical point of view, this inequality was successfully applied to establish new relationships of the generalizations of the indexes ABC and R; in particular, it was applied to prove Theorem 4 and Corollary 3. In addition, other new relationships were obtained between these indices (Theorems 1, 2, and 5) that generalized and improved already known results.

Author Contributions: Investigation, P.B., E.D.M., J.M.R. and J.M.S. All authors have read and agreed to the published version of the manuscript.

Funding: This research was supported by a grant from Agencia Estatal de Investigación (PID2019-106433GBI00/ AEI/10.13039/501100011033), Spain. The research of José M. Rodríguez was supported by the Madrid Government (Comunidad de Madrid-Spain) under the Multiannual Agreement with UC3M in the line of the Excellence of University Professors (EPUC3M23) and in the context of the V PRICIT (Regional Programme of Research and Technological Innovation).

Institutional Review Board Statement: Not applicable.

Informed Consent Statement: Not applicable.

Data Availability Statement: Not applicable.

Acknowledgments: We would like to thank the reviewers for their careful reading of the manuscript and their suggestions, which have improved the presentation of this work.

Conflicts of Interest: The authors declare no conflict of interest.

References

1. Molina, E.D.; Rodríguez, J.M.; Sánchez, J.L.; Sigarreta, J.M. Some Properties of the Arithmetic–Geometric Index. *Symmetry* **2021**, *13*, 857. [CrossRef]
2. Devillers, J.; Balaban, A.T. (Eds.) *Topological Indices and Related Descriptors in QSAR and QSPR*; Gordon and Breach: Amsterdam, The Netherlands, 1999.
3. Karelson, M. *Molecular Descriptors in QSAR/QSPR*; Wiley-Interscience: New York, NY, USA, 2000.

4. Todeschini, R.; Consonni, V. *Handbook of Molecular Descriptors*; Wiley-VCH: Weinheim, Germany, 2000.
5. Gutman, I.; Furtula, B.; Katanić, V. Randić index and information. *AKCE Int. J. Graphs Comb.* **2018**, *15*, 307–312. [CrossRef]
6. Estrada, E. Quantifying network heterogeneity. *Phys Rev. E* **2010**, *82*, 066102. [CrossRef] [PubMed]
7. Pineda, J.; Martínez, C.; Mendez, A.; Muños, J.; Sigarreta, J.M. Application of Bipartite Networks to the Study of Water Quality. *Sustainability* **2020**, *12*, 5143. [CrossRef]
8. Estrada, E.; Torres, L.; Rodríguez, L.; Gutman, I. An atom-bond connectivity index: Modelling the enthalpy of formation of alkanes. *Indian J. Chem.* **1998**, *37A*, 849–855.
9. Furtula, B.; Graovac, A.; Vukicević, D. Augmented Zagreb index. *J. Math. Chem.* **2010**, *48*, 370–380. [CrossRef]
10. Chen, X.; Das, K.C. Solution to a conjecture on the maximum ABC index of graphs with given chromatic number. *Discr. Appl. Math.* **2018**, *251*, 126–134. [CrossRef]
11. Chen, X.; Hao, G. Extremal graphs with respect to generalized ABC index. *Discr. Appl. Math.* **2018**, *243*, 115–124. [CrossRef]
12. Das, K.C.; Elumalai, S.; Gutman, I. On ABC Index of Graphs. *MATCH Commun. Math. Comput. Chem.* **2017**, *78*, 459–468.
13. Das, K.C.; Gutman, I.; Furtula, B. On atom-bond connectivity index. *Chem. Phys. Lett.* **2011**, *511*, 452–454. [CrossRef]
14. Das, K.C.; Rodríguez, J.M.; Sigarreta, J.M. On the maximal general ABC index of graphs with given maximum degree. *Appl. Math. Comput.* **2020**, *386*, 125531. [CrossRef]
15. Gutman, I.; Furtula, B.; Ivanović, M. Notes on Trees with Minimal Atom–Bond Connectivity Index. *MATCH Commun. Math. Comput. Chem.* **2012**, *67*, 467–482.
16. Bollobás, B.; Erdős, P. Graphs of extremal weights. *Ars Combin.* **1998**, *50*, 225–233. [CrossRef]
17. Miličević, A.; Nikolić, S. On variable Zagreb indices. *Croat. Chem. Acta* **2004**, *77*, 97–101.
18. Li, X.; Gutman, I. *Mathematical Aspects of Randić Type Molecular Structure Descriptors*; Univ. Kragujevac: Kragujevac, Serbia, 2006.
19. Nikolić, S.; Kovačević, G.; Miličević, A.; Trinajstić, N. The Zagreb indices 30 years after. *Croat. Chem. Acta* **2003**, *76*, 113–124.
20. Randić, M. On characterization of molecular branching. *J. Am. Chem. Soc.* **1975**, *97*, 6609–6615. [CrossRef]
21. Hua, H.; Das, K.C.; Wang, H. On atom-bond connectivity index of graphs. *J. Math. Anal. Appl.* **2019**, *479*, 1099–1114. [CrossRef]
22. Martínez-Pérez, A.; Rodríguez, J.M.; Sigarreta, J.M. A new approximation to the geometric-arithmetic index. *J. Math. Chem.* **2018**, *56*, 1865–1883. [CrossRef]
23. Rodríguez, J.M.; Sánchez, J.L.; Sigarreta, J.M. Inequalities on the inverse degree index. *J. Math. Chem.* **2019**, *57*, 1524–1542. [CrossRef]
24. Hollas, B. On the variance of topological indices that depend on the degree of a vertex. *MATCH Commun. Math. Comput. Chem.* **2005**, *54*, 341–350.

Article

Local Antimagic Chromatic Number for Copies of Graphs

Martin Bača [1,†], Andrea Semaničová-Feňovčíková [1,*,†] and Tao-Ming Wang [2,†]

1. Department of Applied Mathematics and Informatics, Technical University, 042 00 Košice, Slovakia; martin.baca@tuke.sk
2. Department of Applied Mathematics, Tunghai University, Taichung 40704, Taiwan; wang@go.thu.edu.tw
* Correspondence: andrea.fenovcikova@tuke.sk
† These authors contributed equally to this work.

Abstract: An edge labeling of a graph $G = (V, E)$ using every label from the set $\{1, 2, \ldots, |E(G)|\}$ exactly once is a *local antimagic labeling* if the vertex-weights are distinct for every pair of neighboring vertices, where a vertex-weight is the sum of labels of all edges incident with that vertex. Any *local antimagic labeling* induces a proper vertex coloring of G where the color of a vertex is its vertex-weight. This naturally leads to the concept of a local antimagic chromatic number. The local antimagic chromatic number is defined to be the minimum number of colors taken over all colorings of G induced by *local antimagic labelings* of G. In this paper, we estimate the bounds of the local antimagic chromatic number for disjoint union of multiple copies of a graph.

Keywords: *local antimagic labeling*; local antimagic chromatic number; copies of graphs

MSC: 05C78; 05C69

1. Introduction

In this paper, we will consider only finite graphs without loops or multiple edges. For graph theoretic terminology we refer to the book by Chartrand and Lesniak [1].

An antimagic labeling of a graph $G = (V, E)$ is a bijection f from the set of edges of G to the integers $\{1, 2, \ldots, |E(G)|\}$ such that all vertex-weights are pairwise distinct, where a vertex-weight is the sum of labels of all edges incident with that vertex, i.e., for the vertex $u \in V(G)$ the weight $wt(u) = \sum_{uv \in E(G)} f(uv)$. A graph is called *antimagic* if it admits an antimagic labeling.

The concept of *antimagic labeling* was introduced by Hartsfield and Ringel [2] who conjectured that every simple connected graph, other than K_2, is antimagic. This conjecture is still open although for some special classes of graphs it was proved, see for instance [3–8]. Alon et al. [9] proved that large dense graphs are antimagic. Hefetz et al. [10] proved that any graph on p^k vertices that admits a C_p-factor, where p is an odd prime and k is a positive integer, is antimagic. Perhaps the most remarkable result to date is the proof for regular graphs of odd degree given by Cranston et al. in [11], which was subsequently adapted to regular graphs of even degree by Bércz et al. in [12] and by Chang et al. in [13].

Recently, two groups of authors in [14,15] independently introduced a *local antimagic labeling* as local version of the Hartsfield and Ringel's concept of antimagic labeling. An edge labeling using every label from the set $\{1, 2, \ldots, |E(G)|\}$ exactly once is a *local antimagic labeling* if the vertex-weights $wt(u)$ and $wt(v)$ are distinct for every pair of neighboring vertices u, v.

In [14] authors conjectured that any connected graph other than K_2 admits a *local antimagic labeling*. Bensmail et al. [15] propose the slightly stronger form of the previous conjecture that every graph without component isomorphic to K_2 has a *local antimagic labeling*. This conjecture was proved by Haslegrave [16] using the probabilistic method.

Any *local antimagic labeling* induces a proper vertex coloring of G where the vertex-weight $wt(u)$ is the color of u. This naturally leads to the concept of a local antimagic chromatic number introduced in [14]. The *local antimagic chromatic number* $\chi_{la}(G)$ is defined to be the minimum number of colors taken over all colorings of G induced by *local antimagic labelings* of G.

For any graph G, $\chi_{la}(G) \geq \chi(G)$, where $\chi(G)$ is the chromatic number of G as the minimum number of colors needed to produce a proper coloring of G. In [14] is investigated the local antimagic chromatic number for paths, cycles, friendship graphs, wheels and complete bipartite graphs. Moreover, there is proved that for any tree T with l leaves $\chi_{la}(T) \geq l + 1$.

In this paper, we investigate the local antimagic chromatic number for disjoint union of multiple copies of a graph G, denoted by mG, $m \geq 1$, and we present some estimations of this parameter.

Please note that G does not have to be necessarily connected. By the symbol x_i we denote the element (vertex or edge) corresponding to the element (vertex or edge) x in the ith copy of G in mG, $i = 1, 2, \ldots, m$.

2. Graphs with Vertices of Even Degrees

A graph G is called *equally 2-edge colorable* if it is possible to color its edges with two colors c_1, c_2 such that for every vertex $v \in V(G)$ the number of edges incident to the vertex v colored with color c_1 is the same as the number of edges incident to the vertex v colored with color c_2. This means that for any vertex $v \in V(G)$ is $n^1(v) = n^2(v)$, where $n^i(v)$ denotes the number of edges incident to the vertex v and colored with color c_i, $i = 1, 2$. Trivially, if a graph G is equally 2-edge colorable then all vertices in G have even degrees.

Consider that G is an even regular graph. Then there exists an Euler circle in G. If we alternatively color the edges in the Euler circle with colors c_1 and c_2 we obtain that either for every vertex v in G holds $n^1(v) = n^2(v)$, or there exists exactly one vertex in G, say w, such that $n^1(w) = n^2(w) + 2$.

Consider a 2-edge coloring c of a graph G. Let $c(G)$ denote the number of vertices in G such that $n_c^1(v) \neq n_c^2(v)$ under the labeling c. In this case we say that c is a $c(G)$-*equally 2-edge coloring* of G.

Let c be any 2-edge coloring of G. Let f be any bijective mapping in G, $f : E(G) \to \{1, 2, \ldots, |E(G)|\}$. We define an edge labeling g of mG, $m \geq 1$ in the following way

$$g(e_i) = \begin{cases} m(f(e) - 1) + i, & \text{if } c(e) = c_1 \text{ and } i = 1, 2, \ldots, m, \\ mf(e) + 1 - i, & \text{if } c(e) = c_2 \text{ and } i = 1, 2, \ldots, m. \end{cases}$$

If an edge in G is labeled with the number t, $1 \leq t \leq |E(G)|$, then the corresponding edges in mG are labeled with numbers from the set $\{m(t-1)+1, m(t-1)+2, \ldots, mt\}$. Thus we immediately obtain that the labeling g is a bijective mapping that assigns numbers $1, 2, \ldots, m|E(G)|$ to the edges of mG.

Moreover, for the weight of the vertex v_i, $i = 1, 2, \ldots, m$, in mG under the labeling g we obtain the following

$$
\begin{aligned}
wt_g(v_i) &= \sum_{uv \in E(G)} g(u_i v_i) = \sum_{uv \in E(G):\, c(uv) = c_1} g(u_i v_i) + \sum_{uv \in E(G):\, c(uv) = c_2} g(u_i v_i) \\
&= \sum_{uv \in E(G):\, c(uv) = c_1} (m(f(uv) - 1) + i) + \sum_{uv \in E(G):\, c(uv) = c_2} (mf(uv) + 1 - i) \\
&= m \sum_{uv \in E(G):\, c(uv) = c_1} f(uv) + (i - m) n_c^1(v) \\
&\quad + m \sum_{uv \in E(G):\, c(uv) = c_2} f(uv) + (1 - i) n_c^2(v) \\
&= m \sum_{uv \in E(G)} f(uv) + (i - m) n_c^1(v) + (1 - i) n_c^2(v)
\end{aligned}
$$

$$=m\cdot wt_f(v)+(i-m)n_c^1(v)+(1-i)n_c^2(v).$$

Thus, for every vertex $v \in V(G)$ such that $n_c^1(v) = n_c^2(v) = \deg(v)/2$ we obtain

$$wt_g(v_i) = m \cdot wt_f(v) + \frac{(1-m)\deg(v)}{2} \quad (1)$$

for $i = 1, 2, \ldots, m$. This means that the corresponding vertices in different copies have the same weights. Summarizing the previous we obtain the following lemma that will be used later.

Lemma 1. *Let G be a graph and let c be a 2-edge coloring of G and let f, $f : E(G) \to \{1, 2, \ldots, |E(G)|\}$, be a bijection. Let m, $m \geq 1$, be a positive integer. Then there exists an edge labeling g of mG such that the weights of vertices v_i, $i = 1, 2, \ldots, m$, corresponding to the vertex $v \in V(G)$ satisfying $n_c^1(v) = n_c^2(v) = \deg(v)/2$ will be the same.*

Immediately from the previous result we obtain the following theorem for equally 2-edge colorable graphs.

Theorem 1. *Let m be a positive integer. Let G be an equally 2-edge colorable graph and let f be a local vertex antimagic edge labeling of G that uses $\chi_{la}(G)$ colors. Let for every edge $uv \in E(G)$ be*

$$mwt_f(v) + \tfrac{(1-m)\deg(v)}{2} \neq mwt_f(u) + \tfrac{(1-m)\deg(u)}{2}.$$

Then

$$\chi_{la}(mG) \leq \chi_{la}(G).$$

Proof. Let f be a local vertex antimagic edge labeling of G that uses $\chi_{la}(G)$ colors. Let c be an equally 2-edge coloring of G. This means that for every vertex $v \in V(G)$ is $n^1(v) = n^2(v) = \deg(v)/2$.
According to Lemma 1 and Equality (1) we obtain that there exists a labeling g of mG, $m \geq 1$, such that for every $v \in V(G)$ and every $i = 1, 2, \ldots, m$ holds $wt_g(v_i) = m \cdot wt_f(v) + (1-m)\deg(v)/2$. Thus, g is such labeling that the corresponding vertices in different copies have the same weights. If for all adjacent vertices $u, v \in V(G)$ holds

$$mwt_f(v) + \tfrac{(1-m)\deg(v)}{2} \neq mwt_f(u) + \tfrac{(1-m)\deg(u)}{2} \quad (2)$$

then also all adjacent vertices in mG have distinct weights.
Moreover, $\chi_{la}(mG) \leq \chi_{la}(G)$. This concludes the proof. □

Note, if G is a regular graph then the condition (2) trivially holds. Results in the next two theorems are based on the Petersen Theorem.

Proposition 1. (Petersen Theorem) *Let G be a $2r$-regular graph. Then there exists a 2-factor in G.*

Notice that after removing edges of the 2-factor guaranteed by Petersen Theorem we have again an even regular graph. Thus, by induction, an even regular graph has a 2-factorization.

Theorem 2. *Let G be a $4r$-regular graph, $r \geq 1$. Then for every positive integer m*

$$\chi_{la}(mG) \leq \chi_{la}(G).$$

Proof. Let G be a $4r$-regular graph. According to Petersen Theorem G is decomposable into 2-factors $F_1, F_2, \ldots, F_{2r}$. Consider an edge coloring c of G defined such that

$$c(e) = \begin{cases} c_1, & \text{if } e \in E(F_j), j = 1, 2, \ldots, r, \\ c_2, & \text{if } e \in E(F_j), j = r+1, r+2, \ldots, 2r. \end{cases}$$

Evidently, c is an equally 2-edge coloring of G. Thus, immediately according to Theorem 1 we obtain the desired result. □

Theorem 3. *Let G be a $(4r + 2)$-regular graph, $r \geq 0$, containing a 2-factor consisting only from even cycles. Then for every positive integer m*

$$\chi_{la}(mG) \leq \chi_{la}(G).$$

Proof. Let G be a $(4r + 2)$-regular graph containing a 2-factor consisting only from even cycles. Denote this 2-factor by F_1. Let us denote the edges in component F_1 by the symbols $e_1, e_2, \ldots, e_{|VG|}$ arbitrarily in such a way that all cycles in F_1 are of the form $e_s e_{s+1} e_{s+2} \cdots e_{s+t}$, where s, t are odd integers. As all cycles in F_1 are even, evidently every vertex in G is incident with an edge in F_1 with an even and also with an odd index. According to Petersen Theorem the graph $G - F_1$ is decomposable into 2-factors $F_2, F_3, \ldots, F_{2r+1}$. Consider an edge coloring c of G defined such that

$$c(e) = \begin{cases} c_1, & \text{if } e \in E(F_1), e = e_{2i-1}, i = 1, 2, \ldots, \frac{|V(G)|}{2}, \\ & \text{or if } e \in E(F_j), j = 2, 3, \ldots, r+1, \\ c_2, & \text{if } e \in E(F_1), e = e_{2i}, i = 1, 2, \ldots, \frac{|V(G)|}{2}, \\ & \text{or if } e \in E(F_j), j = r+2, r+3, \ldots, 2r+1. \end{cases}$$

It is easy to see that for every vertex $v \in V(G)$ holds

$$n^1(v) = n^2(v) = 2r + 1.$$

This means that c is an equally 2-edge coloring of G. By Theorem 1 we obtain that $\chi_{la}(mG) \leq \chi_{la}(G)$. □

Corollary 1. *Let n, m be positive integers, $n \geq 2, m \geq 1$. Then*

$$\chi_{la}(mC_{2n}) = 3.$$

Proof. In [14] it was proved that $\chi_{la}(C_k) = 3$ for every $k \geq 3$. According to Theorem 3 we obtain that if $k = 2n$ then for every positive integer m holds $\chi_{la}(mC_{2n}) \leq 3$.
Now suppose there exists a local antimagic labeling f that induces a 2-coloring $\mathcal{C}$ of mC_{2n}, i.e., the set of the vertex weights consists of two numbers $\mathcal{C}_1$ and $\mathcal{C}_2$. As every edge label contributes exactly once to the vertex weight of a vertex colored $\mathcal{C}_1$ we obtain

$$mn \cdot \mathcal{C}_1 = 1 + 2 + \cdots + 2nm.$$

However, every edge label contributes also exactly once to the vertex weight of a vertex colored $\mathcal{C}_2$ thus

$$mn \cdot \mathcal{C}_2 = 1 + 2 + \cdots + 2nm.$$

A contradiction. Thus, $\chi_{la}(mC_{2n}) \geq 3$. □

Theorem 4. *Let n, m be positive integers, $n \geq 1, m \geq 1$. Then*

$$\chi_{la}(mC_{2n+1}) \leq m + 2.$$

Proof. Let us denote the vertex set and the edge set of mC_{2n+1} such that $V(C_{2n+1}) = \{v_i^j : i = 1, 2, \ldots, 2n+1, j = 1, 2, \ldots, m\}$ and $E(C_{2n+1}) = \{v_i^j v_{i+1}^j : i = 1, 2, \ldots, 2n, j = 1, 2, \ldots, m\} \cup \{v_1^j v_{2n+1}^j : j = 1, 2, \ldots, m\}$. Let $e_i^j = v_i^j v_{i+1}^j, i = 1, 2, \ldots, 2n, j = 1, 2, \ldots, m$ and let $e_{2n+1}^j = v_1^j v_{2n+1}^j, j = 1, 2, \ldots, m$.

We define an edge labeling f of mC_{2n+1} in the following way

$$f(e_i^j) = \begin{cases} \frac{m(i-1)}{2} + j, & \text{if } i = 1, 3, \ldots, 2n+1, j = 1, 2, \ldots, m, \\ m\left(2n + 2 - \frac{i}{2}\right) + 1 - j, & \text{if } i = 2, 4, \ldots, 2n, j = 1, 2, \ldots, m. \end{cases}$$

For the weight of the vertex v_i^j, $i = 3, 5, \ldots, 2n+1$, $j = 1, 2, \ldots, m$ we obtain

$$wt_f(v_i^j) = f(e_{i-1}^j) + f(e_i^j) = [m\left(2n + 2 - \frac{i-1}{2}\right) + 1 - j] + [\frac{m(i-1)}{2} + j] = m(2n+2) + 1$$

and for $i = 2, 4, \ldots, 2n$, $j = 1, 2, \ldots, m$, we obtain

$$wt_f(v_i^j) = f(e_{i-1}^j) + f(e_i^j) = [\frac{m((i-1)-1)}{2} + j] + [m\left(2n + 2 - \frac{i}{2}\right) + 1 - j]$$
$$= m(2n+1) + 1.$$

The weight of the vertex v_1^j, $j = 1, 2, \ldots, m$, is

$$wt_f(v_1^j) = f(e_1^j) + f(e_{2n+1}^j) = [\frac{m(1-1)}{2} + j] + [\frac{m((2n+1)-1)}{2} + j] = mn + 2j,$$

thus the weights are $mn + 2, mn + 4, \ldots, m(n+2)$. Thus, all adjacent vertices have distinct weights. Moreover we obtain $\chi_{la}(mC_{2n+1}) \leq m + 2$. □

Please note that a cycle C_{2n+1} is 1-equally 2-edge colorable. It is possible to generalize the results from the previous section also for $c(G)$-equally 2-edge colorable graphs. If we are able to guarantee that for every edge $uv \in E(G)$ is

$$mwt_f(u) + (i - m)n_c^1(u) + (1 - i)n_c^2(u)$$
$$\neq mwt_f(v) + (i - m)n_c^1(v) + (1 - i)n_c^2(v) \qquad (3)$$

then we can prove that

$$\chi_{la}(mG) \leq \chi_{la}(G) + \min\{(m-1)c(G) : c \text{ is a 2-edge coloring of } G \text{ satisfying (3)}\}.$$

This condition is fulfilled for some graphs containing pendant vertices, thus also for some trees.

Lemma 2. *Let G be a graph with l leaves, $l \geq 0$. Then*

$$\chi_{la}(G) \geq l + 1.$$

Proof. The proof is similar to the proof in [14]. Let f be any *local antimagic labeling* of a graph G. Then in the coloring induced by f, the color of a leaf v is $f(uv)$, where $uv \in E(G)$. Hence all the leaves receive distinct colors. Moreover, for any non-leaf w incident with an edge e with $f(e) = |E(G)|$, the color assigned to w is larger than $|E(G)|$. Hence the number of colors in the coloring induced by f is at least $l + 1$. □

Theorem 5. *Let G be a graph without a component isomorphic to K_2 such that all vertices in G but leaves have the same even degree. If there exists a 2-edge coloring c of G such that for all vertices v but leaves holds $n_c^1(v) = n_c^2(v) = \deg(v)/2$, then*

$$ml + 1 \leq \chi_{la}(mG) \leq \chi_{la}(G) + (m-1)l,$$

where m is a positive integer and l is the number of leaves in G.

Proof. Let G be a graph without a component isomorphic to K_2 such that all its vertices but leaves have the same even degree $2r$. Let c be a 2-edge coloring of G such that for all vertices v in G but leaves holds $n_c^1(v) = n_c^2(v) = \deg(v)/2 = r$.
Let f be any local antimagic labeling of a graph G that uses $\chi_{la}(G)$ colors. Then using Equality (1) we obtain that there exists an edge labeling g of mG, $m \geq 1$, such that the weights of non-leaf vertices v_i, $i = 1, 2, \ldots, m$, corresponding to a vertex v in G, are

$$wt_g(v_i) = m \cdot wt_f(v) + (1-m)r.$$

This means that the weights of corresponding non-leaf vertices in every copy of G are the same. However, this also means that the adjacent non-leaf vertices in mG have distinct weights.

Now consider the edges $w_i u_i$, $i = 1, 2, \ldots, m$, where w is a leaf. For $i = 1, 2, \ldots, m$ trivially holds

$$wt_g(w_i) = g(w_i u_i) < \sum_{uv \in E(G)} g(v_i u_i) = wt_g(u_i).$$

Which means that all adjacent vertices have distinct weights.
Combining the previous arguments we obtain

$$\chi_{la}(mG) \leq \chi_{la}(G) + (m-1)l.$$

The lower bound for $\chi_{la}(mG)$ follows from Lemma 2. □

3. Trees

If the graph in Theorem 5 is a forest we immediately obtain the following result.

Theorem 6. *Let T be a forest with no component isomorphic to K_2 such that all vertices but leaves have the same even degree. Then*

$$ml + 1 \leq \chi_{la}(mT) \leq \chi_{la}(T) + (m-1)l,$$

where m is a positive integer and l is the number of leaves in T.

Proof. Trivially, any graph containing K_2 as a component cannot be local antimagic.
Let T be a forest with no component isomorphic to K_2 such that all vertices but leaves have the same even degree $2r$. Clearly there exists a 2-edge coloring c of T such that for all vertices v but leaves hold $n_c^1(v) = n_c^2(v) = \deg(v)/2 = r$. Thus, according to Theorem 5 we are done. □

Immediately from the previous theorem we obtain the result for copies of paths and copies of some stars as $\chi_{la}(P_n) = 3$ for $n \geq 3$ and $\chi_{la}(K_{1,n}) = n + 1$ for $n \geq 2$, see [14].

Corollary 2. *Let P_n be a path on n vertices, $n \geq 3$. Then for every positive integer m, $m \geq 1$, holds*

$$\chi_{la}(mP_n) = 2m + 1.$$

Corollary 3. *Let $K_{1,2n}$ be a star, $n \geq 1$. Then for every positive integer m, $m \geq 1$, holds*

$$\chi_{la}(mK_{1,2n}) = 2nm + 1.$$

Theorem 7. *Let $K_{1,2n+1}$ be a star, $n \geq 1$. Then for every positive integer m, $m \geq 1$, holds*

$$\chi_{la}(mK_{1,2n+1}) = \begin{cases} (2n+1)m + 1, & \text{if } m \text{ is odd or if } m \text{ is even and } m \geq n+1, \\ (2n+1)m + 2, & \text{if } m \text{ is even and } m < n+1. \end{cases}$$

Proof. Let us denote the vertices and the edges of $mK_{1,2n+1}$ such that

$$V(mK_{1,2n+1}) = \{w_i, v_i^j : i = 1, 2, \ldots, m, j = 1, 2, \ldots, 2n+1\},$$
$$E(mK_{1,2n+1}) = \{w_i v_i^j : i = 1, 2, \ldots, m, j = 1, 2, \ldots, 2n+1\}.$$

We consider two cases according to the parity of m.

Case 1: when m is odd.

We define an edge labeling g of $mK_{1,2n+1}$ in the following way

$$g(w_i v_i^j) = \begin{cases} i, & \text{if } j = 1 \text{ and } i = 1, 2, \ldots, m, \\ \frac{3m+1}{2} + i, & \text{if } j = 2 \text{ and } i = 1, 2, \ldots, \frac{m-1}{2}, \\ \frac{m+1}{2} + i, & \text{if } j = 2 \text{ and } i = \frac{m+1}{2}, \frac{m+3}{2}, \ldots, m, \\ 3m + 1 - 2i, & \text{if } j = 3 \text{ and } i = 1, 2, \ldots, \frac{m-1}{2}, \\ 4m + 1 - 2i, & \text{if } j = 3 \text{ and } i = \frac{m+1}{2}, \frac{m+3}{2}, \ldots, m, \\ (j-1)m + i, & \text{if } j = 4, 5, \ldots, n+2 \text{ and } i = 1, 2, \ldots, m, \\ jm + 1 - i, & \text{if } j = n+3, n+4, \ldots, 2n+1 \text{ and } i = 1, 2, \ldots, m. \end{cases}$$

Evidently g is a bijection and the induced weights of the vertices w_i, $i = 1, 2, \ldots, m$, are

$$wt_g(w_i) = \sum_{j=1}^{2n+1} g(w_i v_i^j) = \frac{(2n+1)(m(2n+1)+1)}{2}.$$

As all vertices of degree $2n+1$ have the same weights and the weights of the leaves are distinct we obtain $\chi_{la}(mK_{1,2n+1}) \leq (2n+1)m + 1$. The lower bound follows from Lemma 2.

Case 2: when m is even.

In this case consider a labeling f of $mK_{1,2n+1}$ defined such that

$$f(w_i v_i^j) = \begin{cases} j, & \text{if } j = 1, 2, \ldots, 2n+1 \text{ and } i = 1, \\ 2n + 1 + g(w_{i-1} v_{i-1}^j), & \text{if } j = 1, 2, \ldots, 2n+1 \text{ and } i = 2, 3, \ldots, m. \end{cases}$$

According to the properties of the labeling g, the labeling f is a bijective mapping that assigns numbers $1, 2, \ldots, m(2n+1)$ to the edges of $mK_{1,2n+1}$. The weights of vertices w_i, $i = 2, 3, \ldots, m$, are all the same as

$$wt_f(w_i) = \sum_{j=1}^{2n+1} f(w_i v_i^j) = \sum_{j=1}^{2n+1} \left[2n + 1 + g(w_{i-1} v_{i-1}^j)\right] = (2n+1)^2 + \frac{(2n+1)(m(2n+1)+1)}{2}.$$

The weight of the vertex w_1 is

$$wt_f(w_1) = \sum_{j=1}^{2n+1} f(w_1 v_1^j) = \sum_{j=1}^{2n+1} j = (n+1)(2n+1).$$

If the weight of the vertex w_1 under the labeling f is the same as the weight of some leaf, we obtain that $\chi_{la}(mK_{1,2n+1}) \leq (2n+1)m + 1$. This is satisfied when $(n+1)(2n+1) \leq m(2n+1)$, that is if $n + 1 \leq m$. The equality $\chi_{la}(mK_{1,2n+1}) = (2n+1)m + 1$ holds because the number of induced colors must be greater then the number of leaves, see Lemma 2.

Now consider that the weight of the vertex w_1 under the labeling f is greater then the weight of all leaves, i.e., $n + 1 > m$. Then labeling f induces $(2n+1)m + 2$ colors for vertices. To prove that it is not possible to obtain $(2n+1)m + 1$ colors it is sufficient to consider the fact, that the weight of any vertex of degree $2n + 1$ is at least the sum of numbers $1, 2, \ldots, 2n+1$, thus it is at least $(n+1)(2n+1)$. However, the weights of leaves are at most $(2n+1)m$. Thus if there exists an edge labeling that induces $(2n+1)m + 1$ colors for vertices, under this labeling all vertices w_i, $i = 1, 2, \ldots, m$ must have the same color/weight, say $c(w)$. However, in this case the sum of all edge labels must be equal to m multiple of $c(w)$, as every edge label contributes exactly once the weight of a vertex of degree $2n+1$. Thus $mc(w) = 1 + 2 + \cdots + (2n+1)m$ which implies

$$2c(w) = (2n+1)((2n+1)m+1).$$

However, this is a contradiction as for m even the right side of the previous equation is odd. This means that in this case $\chi_{la}(mK_{1,2n+1}) = (2n+1)m + 2$. □

Please note that Theorem 6 can be extended also for other trees (forests) such that their non-leaf vertices have even degrees, not necessarily the same. We just need to guarantee that the adjacent non-leaf vertices will have distinct weights. For some trees, for example for spiders, we are able to do it. A *spider graph* is a tree with exactly one vertex of degree greater than 2. By $S(n_1, n_2, \ldots, n_l)$, $1 \leq n_i \leq n_{i+1}$, $i = 1, 2, \ldots, l-1$, $l \geq 3$, we denote a spider obtained by identifying one leaf in paths P_{n_i+1}, $i = 1, 2, \ldots, l$. In [17] was proved that if $n_1 = 1$ then $\chi_{la}(S(n_1, n_2, \ldots, n_l)) = l+1$ and if $n_1 \geq 2$ then $\chi_{la}(S(n_1, n_2, \ldots, n_l)) \leq l+2$. Moreover, for $l \geq 4$ the described edge labeling induces for the root vertex, the vertex of degree l, the largest weight over all other vertex weights. Using the presented results we obtain

Theorem 8. *Let $S(n_1, n_2, \ldots, n_l)$ be a spider graph. If l is even, $l \geq 4$, and $n_1 = 1$*

$$\chi_{la}(mS(n_1, n_2, \ldots, n_l)) = ml + 1.$$

If l is even, $l \geq 4$, and $n_1 \geq 2$

$$ml + 1 \leq \chi_{la}(mS(n_1, n_2, \ldots, n_l)) \leq ml + 2.$$

In [18] was proposed the following conjecture.

Theorem 9. *Ref. [18] Let T be a tree other than K_2 with l leaves. Then*

$$l + 1 \leq \chi_{la}(T) \leq l + 2.$$

In the light of the previous results trees, for copies of trees we conjecture

Theorem 10. *Let T be a tree other than K_2 with l leaves. Then for every positive integer m, $m \geq 1$,*

$$ml + 1 \leq \chi_{la}(mT) \leq ml + 2.$$

4. Graphs with Chromatic Index 3

In this section we will deal with 3-regular graphs that admit a proper 3-edge coloring.

Theorem 11. *Let G be a 3-regular graph with chromatic index $\chi'(G) = 3$. Then for every odd positive integer m, $m \geq 1$, holds*

$$\chi_{la}(mG) \leq \chi_{la}(G).$$

Proof. Let c be a proper 3-edge coloring of G. Let f be a local vertex antimagic edge labeling of G that uses $\chi_{la}(G)$ colors.
We define a new labeling g of mG, for m odd, in the following way.

$$g(e_i) = \begin{cases} m(f(e) - 1) + i, & \text{if } c(e) = c_1 \text{ and } i = 1, 2, \ldots, m, \\ m(f(e) - 1) + i + \frac{m+1}{2}, & \text{if } c(e) = c_2 \text{ and } i = 1, 2, \ldots, \frac{m-1}{2}, \\ m(f(e) - 1) + i - \frac{m-1}{2}, & \text{if } c(e) = c_2 \text{ and } i = \frac{m+1}{2}, \frac{m+3}{2}, \ldots, m, \\ mf(e) + 1 - 2i, & \text{if } c(e) = c_3 \text{ and } i = 1, 2, \ldots, \frac{m-1}{2}, \\ mf(e) + m + 1 - 2i, & \text{if } c(e) = c_3 \text{ and } i = \frac{m+1}{2}, \frac{m+3}{2}, \ldots, m. \end{cases}$$

It is easy to see that if an edge in G is labeled with the number t, $1 \leq t \leq |E(G)|$, then the corresponding edges in mG are labeled with numbers from the set $\{m(t-1)+1, m(t-1)+2, \ldots, mt\}$. Thus, g is a bijection that assigns numbers $1, 2, \ldots, m|E(G)|$ to the edges of mG. Now we will calculate a vertex weight of the vertex v_i in mG. Let x, y and z be the vertices adjacent to v in G. Without loss of generality we can assume that $c(xv) = c_1$, $c(yv) = c_2$ and $c(zv) = c_3$. Then for $i = 1, 2, \ldots, (m-1)/2$ we obtain

$$\begin{aligned}wt_g(v_i) &= g(x_iv_i) + g(y_iv_i) + g(z_iv_i) \\ &= [m(f(xv)-1)+i] + \left[m(f(yv)-1)+i+\tfrac{m+1}{2}\right] + [mf(zv)+1-2i] \\ &= m(f(xv)+f(yv)+f(zv)) + \tfrac{3-3m}{2} = mwt_f(v) + \tfrac{3-3m}{2}.\end{aligned}$$

If $i = (m+1)/2, (m+3)/2, \ldots, m$ then

$$\begin{aligned}wt_g(v_i) &= g(x_iv_i) + g(y_iv_i) + g(z_iv_i) \\ &= [m(f(xv)-1)+i] + \left[m(f(yv)-1)+i-\tfrac{m-1}{2}\right] + [mf(zv)+m+1-2i] \\ &= m(f(xv)+f(yv)+f(zv)) + \tfrac{3-3m}{2} = mwt_f(v) + \tfrac{3-3m}{2}.\end{aligned} \quad (4)$$

Thus, in all copies the corresponding vertices have the same weights.

Moreover, as the set of weights of vertices in G under the labeling f consists of $\chi_{la}(G)$ distinct numbers we immediately obtain that also the set of weights of vertices in mG under the labeling g consists of $\chi_{la}(G)$ distinct numbers. Thus, $\chi_{la}(mG) \leq \chi_{la}(G)$. □

Analogously, as it was possible to extend the results in Section 2 for graphs with leaves, we can also extend Theorem 11 for some graphs with pendant vertices.

Theorem 12. *Let G be a graph such that all vertices but leaves have degree 3. If there exists a 3-edge coloring c of G such that for all vertices v but leaves hold $n_c^1(v) = n_c^2(v) = n_c^3(v) = 1$, then for every odd positive integer m, $m \geq 1$,*

$$ml + 1 \leq \chi_{la}(mG) \leq \chi_{la}(G) + (m-1)l,$$

where l is the number of leaves in G.

Proof. Let G be a graph such that all its vertices but leaves have degree 3. Let c be a 3-edge coloring of G such that for all vertices v in G but leaves hold $n_c^1(v) = n_c^2(v) = n_c^3(v) = 1$. Let f be any *local antimagic labeling* of a graph G that uses $\chi_{la}(G)$ colors. Then using Equality (4) we obtain that there exists an edge labeling g of mG, m odd $m \geq 1$, such that the weights of non-leaf vertices v_i, $i = 1, 2, \ldots, m$, corresponding to a vertex v in G, are

$$wt_g(v_i) = mwt_f(v) + \tfrac{3-3m}{2}.$$

This means that the weights of corresponding non-leaf vertices in every copy of G are the same. However, this also means that the adjacent non-leaf vertices in mG have distinct weights.

Now consider the edges w_iu_i, $i = 1, 2, \ldots, m$, where w is a leaf. Trivially holds

$$wt_g(w_i) = g(w_iu_i) < \sum_{uv \in E(G)} g(v_iu_i) = wt_g(u_i).$$

Which means that all adjacent vertices have distinct weights.
Combining the previous arguments we obtain

$$\chi_{la}(mG) \leq \chi_{la}(G) + (m-1)l.$$

The lower bound for $\chi_{la}(mG)$ follows from Lemma 2. □

Immediately for forests we obtain the following result.

Corollary 4. *Let T be a forest such that all its vertices but leaves have degree 3. Then for every odd positive integer m, $m \geq 1$ holds*

$$ml + 1 \leq \chi_{la}(mT) \leq \chi_{la}(T) + (m-1)l,$$

where l is the number of leaves in T.

Proof. Let T be a forest such that all its vertices but leaves have degree 3. Trivially there exists a 3-edge coloring c of T such that for all vertices v but leaves hold $n_c^1(v) = n_c^2(v) = n_c^3(v) = 1$. Using Theorem 12 we obtain the desired result. □

The next theorem shows how it is possible to extend the previous result for regular graphs that are decomposable into spanning subgraphs that are all isomorphic either to even regular graphs or 3-regular graphs.

Theorem 13. *Let G be a graph that can be decomposed into factors $G_1, G_2, \ldots, G_k$, $k \geq 1$, and let every factor G_i, $i = 1, 2, \ldots, k$, be isomorphic to a graph of the following types:*

type I: a 4-regular graph,
type II: a 2-regular graph consisting of even cycles,
type III: a 3-regular graph with chromatic index 3.

If every factor G_i, $i = 1, 2, \ldots, k$, is of type I or of type II then for every positive integer m, $m \geq 1$, holds

$$\chi_{la}(mG) \leq \chi_{la}(G).$$

If at least one factor G_i, $i = 1, 2, \ldots, k$, is of type III then for every odd positive integer m, $m \geq 1$, holds

$$\chi_{la}(mG) \leq \chi_{la}(G).$$

Please note that the exact value of $\chi_{la}(K_n)$ is n, since $\chi_{la}(K_n) \geq \chi(K_n) = n$. Immediately from the previous theorem we obtain the following result for complete graphs K_n.

Corollary 5. *Let K_n be a complete graph on n vertices, $n \geq 4$. If $n \equiv 1 \pmod 4$ then for every positive integer m, $m \geq 1$, and if $n \equiv 0 \pmod 4$ then for every odd positive integer m, $m \geq 1$, we have $\chi_{la}(mK_n) = n$.*

5. Conclusions

One interesting problem is to find a local antimagic chromatic number for disjoint union of arbitrary graphs. According to results proved by Haslegrave [16] we obtain that this parameter is finite for disjoint union of arbitrary graphs if and only if non of these graphs contains an isolated edge as a subgraph. Moreover, Haslegrave [16] proved the following result.

Theorem 14. *Ref. [16] For every graph G with m edges, none of which is isolated, and for any positive integer k, the edges of G may be labeled with a permutation of $\{k, k+1, \ldots, m+k-1\}$ in such a way that the vertex sums distinguish all pairs of adjacent vertices.*

Immediately from this result we obtain an upper bound for a local antimagic chromatic number for disjoint union of arbitrary graphs.

Theorem 15. *Let G_i, $i = 1, 2, \ldots, n$, be a graph with no isolated edge. Then*

$$\chi_{la}\left(\bigcup_{i=1}^n G_i\right) \leq \min\left\{\chi_{la}(G_t) + \sum_{i=1}^n |V(G_i)| - |V(G_t)| : t = 1, 2, \ldots, n\right\}.$$

For some graphs we can obtain a better upper bound.

Theorem 16. *Let G_i, $i = 1, 2$, be a graph with no isolated edge. Let G_2 be a graph such that all vertices but leaves have the same degree. Then*

$$\chi_{la}(G_1 \cup G_2) \leq \chi_{la}(G_1) + \chi_{la}(G_2) + l_2,$$

where l_2 is the number of leaves in G_2.

Proof. Let G_2 be a graph such that all vertices but leaves have the same degree r, $r \geq 2$ and let l_2 be the number of leaves in G_2. Let f_i, $i = 1, 2$, be a local vertex antimagic edge labeling of G_i that uses $\chi_{la}(G_i)$ colors. We define an edge labeling g of $G_1 \cup G_2$ such that

$$g(e) = \begin{cases} f_1(e), & \text{if } e \in E(G_1), \\ f_2(e) + |E(G_1)|, & \text{if } e \in E(G_2). \end{cases}$$

As f_1 and f_2 are bijections evidently also g is a bijection. For the vertex weights under the labeling g we obtain the following. If $v \in V(G_1)$ then

$$wt_g(v) = \sum_{uv \in E(G_1)} g(uv) = \sum_{uv \in E(G_1)} f_1(uv) = wt_{f_1}(v).$$

Thus, the weights of adjacent vertices in G_1 are distinct and they induce $\chi_{la}(G_1)$ colors. If $v \in V(G_2)$ and $\deg_{G_2}(v) = r$ then

$$wt_g(v) = \sum_{uv \in E(G_2)} g(uv) = \sum_{uv \in E(G_2)} (f_2(uv) + |E(G_1)|)$$
$$= \sum_{uv \in E(G_2)} f_2(uv) + r|E(G_1)| = wt_{f_2}(v) + r|E(G_1)|.$$

If $v \in V(G_2)$ and $\deg_{G_2}(v) = 1$ then

$$wt_g(v) = \sum_{uv \in E(G_2)} g(uv) = \sum_{uv \in E(G_2)} (f_2(uv) + |E(G_1)|) = f_2(uv) + |E(G_1)|$$
$$= wt_{f_2}(v) + |E(G_1)|.$$

This means that also the weights of adjacent vertices in G_2 are distinct. Moreover, we obtain that the labeling g induces at most $\chi_{la}(G_2) + l_2$ colors as the number of colors assigned to the vertices of degree at least 2 is the same and all the leaves could be assigned with the colors different from the colors of non leaves.

Combining the previous we obtain that the labeling g induces at most $\chi_{la}(G_1) + \chi_{la}(G_2) + l_2$ colors. □

Theorem 17. *Let G be a graph with no isolated edge and with l leaves. Then for every positive integer m, $m \geq 1$ holds*

$$l + 2m + 1 \leq \chi_{la}(G \cup mP_3) \leq \chi_{la}(G) + 2m + 1.$$

Proof. Let G be a graph with no isolated edge and with l leaves. The lower bound follows from Lemma 2. The upper bound is based on the fact that there exists a *local antimagic labeling* of mP_3 that induces $2m + 1$ colors such that the color of every vertex of degree 2 in mP_3 will have the same color $2m + 1$. Thus, the labeling g of $mP_3 \cup G$ described in the proof of Theorem 16 induces also $2m + 1$ colors for vertices in mP_3. These colors are $|E(G)| + 1, |E(G)| + 2, \ldots, |E(G)| + 2m$ and $2|E(G)| + 2m + 1$. In general, these colors are distinct from colors of vertices in G_1 induced by the labeling g. This concludes the proof. □

Author Contributions: Conceptualization, M.B., A.S.-F. and T.-M.W.; methodology, M.B., A.S.-F. and T.-M.W.; validation, M.B., A.S.-F. and T.-M.W.; investigation, M.B., A.S.-F. and T.-M.W.; resources, M.B., A.S.-F. and T.-M.W.; writing—original draft preparation, A.S.-F.; writing—review and editing, M.B., A.S.-F. and T.-M.W.; supervision, A.S.-F.; project administration, M.B., A.S.-F. and T.-M.W.; funding acquisition, M.B., A.S.-F. and T.-M.W. All authors have read and agreed to the published version of the manuscript.

Funding: This work was supported by the Slovak Research and Development Agency under the contract No. APVV-19-0153 and by VEGA 1/0233/18. Also for the author Tao-Ming Wang the research is supported by MOST 108-2115-M-029-002 from the ministry of science and technology of Taiwan.

Institutional Review Board Statement: Not applicable.

Informed Consent Statement: Not applicable.

Data Availability Statement: Not applicable.

Conflicts of Interest: The authors declare no conflict of interest.

References

1. Chartrand, G.; Lesniak, L. *Graphs and Digraphs*, 4th ed.; Chapman and Hall, CRC: Boca Raton, FL, USA, 2005.
2. Hartsfield, N.; Ringel, G. *Pearls in Graph Theory*; Academic Press: Boston, MA, USA, 1994.
3. Cheng, Y. Latice grids and prisms are antimagic. *Theor. Comput. Sci.* **2007**, *374*, 66–73. [CrossRef]
4. Cheng, Y. A new class of antimagic Cartesian product graphs. *Discrete Math.* **2008**, *308*, 6441–6448. [CrossRef]
5. Gallian, J. A dynamic survey of graph labeling. *Electronic J. Combin.* **2017**, *1*, DS6.
6. Shang, J.L.; Lin, C.; Liaw, S.C. On the antimagic labeling of star forests. *Util. Math.* **2015**, *97*, 373–385.
7. Silalahi, R.Y. Antimagic Labelings on Disconnected Graphs. Master's Thesis, National Chung-Hsing University, Taichung, Taiwan, 2019.
8. Wang, T.-M. Toroidal grids are anti-magic. In Proceedings of the 11th Annual International Conference, Kunming, China, 16–19 August 2005; pp. 671–679.
9. Alon, N.; Kaplan, G.; Lev, A.; Roditty, Y.; Yuster, R. Dense graphs are antimagic. *J. Graph Theory* **2004**, *47*, 297–309. [CrossRef]
10. Hefetz, D.; Saluz, A.; Tran, H.T.T. An application of the combinatorial nullstellensatz to a graph labelling problem. *J. Graph Theory* **2010**, *65*, 70–82. [CrossRef]
11. Cranston, D.W.; Liang, Y.C.; Zhu, X. Regular graphs of odd degree are antimagic. *J. Graph Theory* **2015**, *80*, 28–33. [CrossRef]
12. Bércz, K.; Bernáth, A.; Vizer, M. Regular graphs are antimagic. *Electron. J. Comb.* **2015**, *22*, 3. [CrossRef]
13. Chang, F.; Liang, Y.C.; Pan, Z.; Zhu, X. Antimagic labeling of regular graphs. *J. Graph Theory* **2016**, *82*, 339–349. [CrossRef]
14. Arumugam, S.; Premalatha, K.; Bača, M.; Semaničová-Feňovčíková, A. Local antimagic vertex coloring of a graph. *Graphs Comb.* **2017**, *33*, 275–285. [CrossRef]
15. Bensmail, J.; Senhaji, M.; Szabo Lyngsie, K. On a combination of the 1-2-3 conjecture and the antimagic labelling conjecture. *Discrete Math. Theor. Comput. Sci.* **2017**, *19*, 22.
16. Haslegrave, J. Proof of a local antimagic conjecture. *Discrete Math. Theor. Comput. Sci.* **2018**, *20*, 18.
17. Bača, M.; Semaničová-Feňovčíková, A.; Wang, T.-M. Local antimagic chromatic number of some trees. *J. Discret. Math. Sci. Cryptogr.* **2020**. [CrossRef]
18. Arumugam, S.; Lee, Y.-C.; Premalatha, K.; Wang, T.-M. On local antimagic vertex coloring for corona products of graphs. *arXiv* **2018**, arXiv:1808.04956.

Article

Free Cells in Hyperspaces of Graphs

José Ángel Juárez Morales [1], Gerardo Reyna Hernández [1,*], Jesús Romero Valencia [2] and Omar Rosario Cayetano [1]

[1] Faculty of Mathematics, Autonomous University of Guerrero, Carlos E. Adame 54, Col. La Garita, Acapulco 39650, Guerrero, Mexico; 19254713@uagro.mx (J.Á.J.M.); omarrosarioc@gmail.com (O.R.C.)
[2] Faculty of Mathematics, Autonomous University of Guerrero, Sauce 19, La Cima, Chilpancingo de los Bravo 39086, Guerrero, Mexico; 14086@uagro.mx
* Correspondence: 17236@uagro.mx

Abstract: Often for understanding a structure, other closely related structures with the former are associated. An example of this is the study of hyperspaces. In this paper, we give necessary and sufficient conditions for the existence of finitely-dimensional maximal free cells in the hyperspace $C(G)$ of a dendrite G; then, we give necessary and sufficient conditions so that the aforementioned result can be applied when G is a dendroid. Furthermore, we prove that the arc is the unique arcwise connected, compact, and metric space X for which the anchored hyperspace $C_p(X)$ is an arc for some $p \in X$.

Keywords: hyperspace; graph; dendroid; dendrite

1. Introduction

In the study of a mathematical structure, sometimes other structures that allow for visualizing problems in different ways are built.

One of the theories developed using this type of study is the Theory of Hyperspaces; this theory began with the investigations of F. Hausdorff and L. Vietoris. Given a topological space X, the 2^X hyperspace of all nonempty and closed subsets of X was introduced by L. Vietoris in 1922, and he proved basic facts about 2^X—for example, compactness of X implies compactness of 2^X and vice versa; 2^X is connected if and only if X is connected. When X is a metric space, 2^X can be endowed with the Hausdorff metric (defined by F. Hausdorff in 1914).

The hyperspace of all nonempty, closed and connected subsets of X is denoted by $C(X)$ and considered as a subspace of 2^X. In turn, the hyperspace of all nonempty, closed, and connected sets of X containing a point p, which is denoted by $C_p(X)$, is a subspace of $C(X)$.

The hyperspaces $C(X)$ and $C_p(X)$ are subjects of study for many researchers. Among several topics about hyperspace, one of the most interesting is to recognize a hyperspace as homeomorphic to some known space: Ref. [1] presents a special class of spaces X for which $C(X)$ is homeomorphic to the infinite cylinder $X \times \mathbb{R}_{\geq 0}$. Another interesting topic is to analyze topological properties: for compact, connected, and metric X, the hyperspaces $C_p(X)$ are locally connected for all $p \in X$ [2].

Graphs have been widely and deeply studied (see [3–7]) and have proved to be an excellent tool for representing and modeling different structures in several areas of discrete mathematics and computation (see [8,9]). As far as hyperspace is concerned, there exist some works relating both subjects. For example, Duda [10] proved that a space X is a finite graph if and only if $C(X)$ is a polyhedral. In a dendroid X smooth in a point p, $C_p(X)$ is homeomorphic to the Hilbert cube if and only if p is not in the interior of a finite tree in X, a result due to Carl Eberhart [11]. Recently, Reyna et al. proved that, in a local space X, $C_p(X)$ is a polyhedral for all p if and only if X is a finite graph [12].

In this paper, we are concerned with fully determining the existence of maximal finite dimensional free cells in the hyperspace $C(X)$, first of a dendrite and then a dendroid X,

as well as examining necessary and sufficient conditions for the hyperspace $C_p(X)$ if an arc provided X is an arc-wise connected space.

2. Definitions

Throughout this paper, the term *space* is meant to be a connected, compact, and metric space, and a *subspace* is understood to be a subset of a space which is a space itself. Given a space X, the symbol 2^X denotes the *hyperspace* of non-empty closed subsets of X, and $C(X)$ is the hyperspace of non empty subspaces of X both endowed with the *Hausdorff's metric*, two models of hyperspaces are shown in Figure 1. Notice that X is naturally embedded in 2^X via the map $x \mapsto \{x\}$ (compare with ([13] [0.48]).

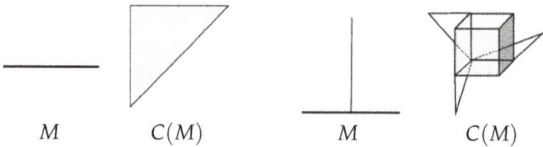

M   C(M)   M   C(M)

Figure 1. The hyperspaces $C(M)$ for the path P_2 and the star S_3.

Given a point $p \in X$, the *anchored hyperspace* of X at p, denoted by $C_p(X)$, is the subspace of $C(X)$ consisting of those elements containing p. Note that $C_p(X)$ is a subspace of $C(X)$, which in turn is a subspace of 2^X.

A space X is *unicoherent* if, for any $A, B \subset X$ subspaces such that $X = A \cup B$, we have that $A \cap B$ is connected. The space X is called *hereditary unicoherent* if each subspace is unicoherent.

A *graph* G, consisting of a set $V(G)$, called the *vertices* of G and a set $E(G)$ of unordered pairs of elements of $V(G)$, called the *edges* of G. Letting G be a graph, if two vertices x and y of G form an edge, we say that they are *adjacent*, and this is denoted by $x \sim y$. This fact is also expressed by saying that x and y are *neighbors*. A vertex of G is called a *ramification vertex* if it has three or more neighbors and a *terminal vertex* if it has exactly one neighbor. G is called *simple* if it contains no loops (a vertex adjacent to itself) and possesses at most one edge between any two vertices. A *path* between two vertices u and v of G is a finite sequence of consecutive adjacent vertices such that the first one is u and the last one is v. G is *connected* if there is a path between any two vertices. A *cycle* in G is a finite sequence of at least three consecutively adjacent vertices such that the first one and the last one are adjacent. In this paper, we consider simple and connected graphs without cycles whose vertices are ramification or terminal vertices, that is, there are no vertices with exactly two neighbors.

In order to consider a graph G as a metric space, if we use the notation $[u, v]$ for the edge joining the vertices u and v, we must identify any edge $[u, v] \in E(G)$ with the closed interval $[0, l]$ (if $l := L([u, v])$; therefore, any point in the interior of any edge is a point of G and, if we consider the edge $[u, v]$ as a graph with just one edge, then it is identified with the closed interval $[0, l]$. A connected graph G is naturally equipped with a distance defined on its points, induced by taking shortest paths in G. Then, we see G as a metric graph (see [10,14]); according to this, a *dendroid* is a simple and connected graph without cycles which is a hereditary unicoherent space; the *comb* and the *harmonic fan* are examples of dendroids (see Figure 2). By *dendrite*, we mean a locally connected dendroid. Any tree, the F_ω, and the *Gehmann* dendrite are examples of these types of graphs (see Figure 3). Throughout this paper, G denotes a dendroid or a dendrite.

Figure 2. The comb and harmonic fan dendroids.

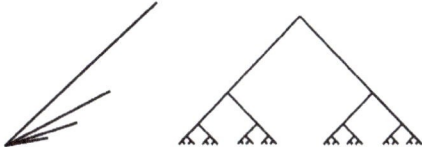

Figure 3. The F_ω and Gehmann dendrites.

A point $p \in G$ is called *essential of type I* if it is a vertex with infinitely many neighbors or *essential of type II* if there exists an infinite sequence of ramification vertices (p_n) such that $p_n \to p$. We use the word *essential* to mean essential of type I or II. A point which is not a vertex, nor an essential point, is called an *ordinary point*; we denote $T(G)$, $O(G)$, $R(G)$, and $ES(G)$ the sets of terminal vertices, ordinary points, ramification vertices, and essential points, respectively.

The *order* of a point x in a dendroid G is defined as follows:

$$o_G(x) = \begin{cases} 1, & \text{if } x \text{ is a terminal vertex;} \\ 2, & \text{if } x \text{ is an ordinary point;} \\ n, & \text{if } x \text{ has exactly } n \text{ neighbors;} \\ \infty, & \text{if } x \text{ is an essential point.} \end{cases}$$

An *m-dimensional cell* (or *m-cell* for short) in a space X is a subspace M homeomorphic to $[0,1]^m$, the part of M homeomorphic to $(0,1)^m$ is called the *interior manifold* of M, and it is denoted by M°, while $M - M^\circ$ is denoted with ∂M, and it is called the *boundary manifold* of M. If it occurs that the interior manifold M° is actually an open set of X, then M is called a *free cell* of X; Figure 4 shows a space with some free cells. In Theorem 2, we establish sufficient and necessary conditions for the existence of a maximal free m-cell in the hyperspace $C(G)$ for a dendrite G. Furthermore, in Theorem 3, we establish sufficient and necessary conditions so that Theorem 2 can be applied when G is an arbitrary dendroid.

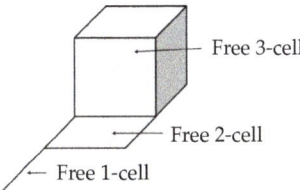

Figure 4. Free cells.

3. Preliminaries

Given $m \leq n$, let A and B be m, n-cells, respectively, with $A \subseteq B$. If $m = n$, A° is an open subset of B°. On the other hand, if $m < n$, then A° is not an open subset of B° because non-empty neighborhoods of B° contain m-dimensional open balls, and none of these can be contained in A°. Therefore, the next lemma follows at once.

Lemma 1. *(a) If a cell is contained in a higher dimensional cell, then the first one is not a free cell.*
(b) Each m-cell contained in a free m-cell is a free m-cell.

In order to show that the cells that we are going to locate in hyperspace $C(G)$ of the dendrite G are maximal, we need Corollary 1 and Lemmas 2–5; in all of these, except Lemma 3, it is assumed that $A \subseteq B$ are m-cells.

Recall that Int A and Bd A designate, respectively, the *topological interior* and *topological boundary* of the set A.

Lemma 2. *If $\partial A = \partial B$, then $A = B$.*

Proof. Since $A^\circ \subseteq B^\circ$, it remains to show that $B^\circ \subseteq A^\circ$. Let $x \in B^\circ$ and suppose $x \notin A^\circ$. Now, if we take $y \in A^\circ$, then $x, y \in B^\circ$. Let α be an arc from x to y contained in B°. Then, the arc α contains an end point in $B - A$ and the other end point in A°. It necessarily occurs that $\alpha \cap \partial A \neq \emptyset$, and this is absurd. □

Recall that the *Borsuk–Ulam* Theorem establishes that, for any continuous map $f : S^n \to \mathbb{R}^n$, there must exist some point $x \in S^n$ such that $f(x) = f(-x)$. This theorem, in particular, implies that no such maps can be *one to one*, and this is the key piece in the proof of next lemma.

Lemma 3. *The unique homeomorphic copy of S^n contained in S^n is S^n itself.*

Proof. Let A be a proper homeomorphic subspace of S^n; notice that we can suppose that the North Pole is not contained in A (otherwise, apply a suitable rotation to S^n). If $\psi : S^n \to A$ is a homeomorphism and σ is the usual stereographic projection of S^n to $\mathbb{R}^n$ restricted to A, then $\sigma \circ \psi : S^n \to \mathbb{R}^n$ is a continuous *one to one* map, a contradiction. □

Corollary 1. *If $\partial A \subseteq \partial B$, then $A = B$.*

Lemma 4. *If $B^\circ \subseteq A$, then $A = B$.*

Proof. Let $x \in \partial B$ and let U be a neighborhood of x in B. Since $U \cap B^\circ \neq \emptyset$, $U \cap A \neq \emptyset$, and hence $x \in \overline{A} = A$. □

Lemma 5. *Let $x \in \partial A$, if $x \notin \partial B$, then $x \in \text{Bd } A$.*

Proof. Suppose $x \notin \text{Bd } A$; then, $x \in \text{Int } A$ and hence Int $A \cap B^\circ$ is an open set in B containing x. Thus, there must exist a neighborhood V of x homeomorphic to $(0,1)^m$ contained in Int $A \cap B^\circ$, and the latter shows that x cannot belong to any face of A. In other words, $x \notin \partial A$. □

If $J = [p_1, p_2]$ is an arc, it is a well known fact that $C(J)$ is a 2-cell whose interior manifold are all subsets in the form $A = [a, b]$, where $a \neq b$ and none of these points equal to p_1 or p_2 (see [10]).

4. Free Cells in Hyperspaces of Dendrites

4.1. The Case $n = 2$

We are close to announcing Theorem 2 where necessary and sufficient conditions are given for the existence of a maximal finite dimensional free n-cells in the hyperspace $C(G)$ of a dendrite G. The free cell built in its proof has the property that all of their elements contain a certain subspace A. In this particular case, the maximal free cells are the hyperspaces $C(J)$ with J an edge, and none of these cells have such a property. Therefore, the case $n = 2$ needs to be treated separately. However, first, it is necessary to state the following known property about locally connected topological spaces.

Lemma 6. *In any locally connected topological space, the components of open sets are open sets.*

Theorem 1. *The hyperspace $C(G)$ of a dendrite G contains a maximal free 2-cell $\mathcal{B}$ if and only if $\mathcal{B} = C(J)$ for some edge J of G.*

Proof. Let J be an edge of G and p_1, p_2 their extremes, consider an element $A = [a,b] \in (C(J))°$ and let $\varepsilon > 0$ such that $N_\varepsilon(A) \subseteq J°$ (where $N_\varepsilon(A)$ is the union of all open balls $B(x, \varepsilon)$ as x ranges over all points of A). Hence, if $B(A, \varepsilon)$ is the open ball (in the Hausdorff metric of $C(G)$) centered at A, then $B(A, \varepsilon) \subseteq (C(J))°$ and $C(J) \subset C(G)$ is a free 2-cell.

Now, we see that the free cell $C(J)$ is maximal. Let $\mathcal{A}$ be a free 2-cell in $C(G)$ containing $C(J)$. If $A \in \partial C(J)$, we have (i) $A = [p_j, x]$, ($j = 1$ or $j = 2$) or else (ii) $A = \{x\}$ with $x \in [p_1, p_2]$. We claim that $A \in \partial \mathcal{A}$. If A is as (i), we have three sub-cases:

(1) *The point p_j is a ramification vertex.* Suppose with no loss of generality that $j = 1$. If $A \notin \partial \mathcal{A}$, then $A \in \mathcal{A}°$ and hence there must exist $\varepsilon > 0$ with $B(A, \varepsilon) \subseteq \mathcal{A}°$. Take $x_1, x_2 \in N_\varepsilon(A) - A$ such that $[x_1, p_1] \cap [x_2, p_1] = \{p_1\}$, $[x_i, p_1] \subseteq B(p_1, \varepsilon)$ ($i \in \{1, 2\}$) and a point $u \in (x, p_2)$ such that $[u, x] \subseteq B(x, \varepsilon)$. The set

$$\mathcal{B} = \{A \cup [w_1, p_1] \cup [w_2, p_1] \cup [x, w_3] \mid w_1 \in [x_1, p_1], w_2 \in [x_2, p_1], w_3 \in [u, x]\}$$

is a 3-cell (see [13] [Theorem 1.100]) contained in $B(A, \varepsilon) \subseteq \mathcal{A}°$, and this is absurd.

(2) *The point p_j is essential.*
A similar analysis as the previous case shows that a 3-cell contained in $B(A, \varepsilon)$ could be built.

(3) *The point p_j is a terminal vertex, and x is an ordinary point or a terminal vertex.* In this case, we have $A \in \text{Int } C(J)$, and this contradicts Lemma 5.

The above shows that $A \in \partial \mathcal{A}$ as desired. For the case ii), if we assume that $A = \{x\}$ and $A \notin \partial \mathcal{A}$, take $H \in \mathcal{A}° - C(J)$ (see Lemma 4) and notice that H does not contain ramification points or essential points; otherwise, in a neighborhood of H contained in $\mathcal{A}°$, 3-cells or even Hilbert cubes can be located (in the proof of Theorem 2, it is shown in detail how is this possible). Hence, H is an arc and let q_1 and q_2 denote their end points; according to this, it must be $p_1 \in [x, q_1]$ or else $p_2 \in [x, q_1]$. Suppose $p_1 \in [x, q_1]$, and, using the fact that $H \neq \{x\}$ and $\mathcal{A}°$ is arcwise connected, take $\alpha \subseteq \mathcal{A}°$ an arc from $\{x\}$ to H. We claim that there exists $L \in \alpha$ such that $p_1 \in L$. Otherwise, we have $\alpha \subseteq C(G) - C_{p_1}(G)$. Let U be the component of $G - \{p_1\}$ containing x, let V be the union of the remaining components and notice that $H \subseteq V$. By Lemma 6, U and V are open sets, hence $\mathcal{U} = \{B \in C(G) \mid B \subseteq U\}$ and $\mathcal{V} = \{B \in C(G) \mid B \subseteq V\}$ are non-empty, disjoint open sets in $C(G) - C_{p_1}(G)$ (compare with [15] [Theorem 4.5]) and therefore the sets $\alpha \cap \mathcal{U}$ and $\alpha \cap \mathcal{V}$ form a separation of α, which is impossible, being α connected. This proves the existence of the desired L.

The point p_1 is a ramification vertex or an essential point; since $L \in \mathcal{A}°$, as in the sub-cases (1) and (2) for some suitable $\varepsilon' > 0$, it is possible to find a 3-cell contained in $B(L, \varepsilon') \subseteq \mathcal{A}°$, and this is a contradiction once again. Therefore, in this case, it must be $A \in \partial \mathcal{A}$ and the result now follows from Corollary 1.

For the converse, let $\mathcal{B} \subseteq C(G)$ be a maximal free 2-cell and let $A \in \mathcal{B}°$. Notice that A does not contain ramification vertices or essential points. The above remarks result in A needing to be an arc; if $J = [p_1, p_2]$ (where p_1, p_2 are vertices of G) is the edge containing A, we claim that $\mathcal{B} \subseteq C(J)$. Otherwise, let $B \in \mathcal{B} - C(J)$. Hence, for each $x \in B$ and for each $y \in A$, it occurs that a) $p_1 \in [x, y]$ or b) $p_2 \in [x, y]$. Suppose without loss of generality that a) occurs and let α be an arc in $\mathcal{B}$ from B to A such that $\alpha - \{B\} \subseteq \mathcal{B}°$. Since $B \in C(J)$ and $A \in C(J)$, there must exist $C \in \alpha \cap \partial C(J)$. Hence, C has the form $[p_1, a]$. If $\varepsilon > 0$ is such that $B(C, \varepsilon) \subseteq \mathcal{B}°$, then $B(C, \varepsilon)$ contains a 3-cell (if p_1 is a ramification point) or even a Hilbert cube (if p_1 is an essential point). This is a contradiction in any case. This shows that $\mathcal{B} \in C(J)$ and therefore $\mathcal{B} = C(J)$. □

4.2. The Case $n > 2$

We need to introduce some terminology about the hyperspace $C(K)$ for a tree K (for more details, see [10,12]).

An *internal tree* T of a tree K is a subgraph which is a tree not containing terminal vertices of K. Let $IT(K)$ denote the set of internal trees of K. For $T \in IT(K)$, let $I_1, \ldots, I_n$ be those edges of K such that $I_i \cap T \neq \emptyset$ and I_i is not contained in T. We define

$$D(1, T) = T \cup \left(\bigcup_{i=1}^{n} I_i \right),$$

and we say that this is the *canonical representation* of $D(1, T)$. Given an internal tree $T \subset K$, let $\mathfrak{M}(T)$ be the family of all subspaces of K in the form

$$((c_i)_{i=1}^{n})_T = T \cup \left(\bigcup_{i=1}^{n} [0_{I_i}, c_i] \right),$$

where 0_{I_i} is the vertex of I_i contained in T, and $[0_{I_i}, c_i]$ is the subarc of I_i joining 0_{I_i} with c_i.

Lemma 7. *Let K be a tree, then*
(i) *For each internal tree $T \subset K$, the family $\mathfrak{M}(T)$ is a n-cell.*
(ii) *The hyperspace of $C(K)$ is*

$$C(K) = \left[\bigcup_{T \in IT(K)} \mathfrak{M}(T) \right] \cup \left[\bigcup_{I \in E(K)} C(I) \right].$$

Theorem 2. *The hyperspace $C(G)$ of a dendrite G contains a maximal free n-cell ($n > 2$) if and only if there exists a tree $K \subseteq G$ satisfying the following conditions:*
(i) $T(K) = \{p_1, \ldots, p_n\} \subseteq R(G) \cup T(G) \cup ES(G)$,
(ii) *for all $x \in K - T(K)$, $o_K(x) = o_G(x)$.*

Proof. For each $p_i \in T(K)$, let $r_i \in R(K)$ such that $[r_i, p_i] \cap R(K) = \{r_i\}$.

Put $A = K - \left(\bigcup_{i=1}^{n} (r_i, p_i] \right)$ and for each $\mathbf{x} = (x_i)_{i=1}^{n} \in \prod_{i=1}^{n} [r_i, p_i]$ let $C_\mathbf{x}$ denote the set $A \cup \left(\bigcup_{i=1}^{n} [r_i, x_i] \right)$. We claim that the family $\mathfrak{M}(A) = \{C_\mathbf{x} \mid \mathbf{x} \in \prod_{i=1}^{n} [r_i, p_i]\}$ is a maximal free n-cell in $C(G)$.

That $\mathfrak{M}(A)$ is actually a n-cell is due to [13], [Theorem 1.100]; therefore, we only need to verify the maximal and free properties.

Let $C_\mathbf{x} \in (\mathfrak{M}(A))^\circ$ and define $L = (G - K) \cup \{p_1, p_2, \ldots, p_n\}$. Put $\alpha = d(C_\mathbf{x}, L) = \inf\{d(c, l) \mid c \in C_\mathbf{x}, l \in L\}$, $\alpha_i = d(x_i, A), \beta_{ij} = d(x_i, [r_j, p_j])$, where $i \neq j$ and $i, j \in \{1, 2, \ldots, n\}$. Since all these quantities are positive, take $\varepsilon > 0$ less than all those and $Y \in B(C_\mathbf{x}, \varepsilon)$. For each $i \in \{1, 2, \ldots, n\}$, choose $z_i \in B(x_i, \varepsilon) \cap Y$ and notice that $z_i \notin A \cup L \cup [r_j, p_j]$ if $i \neq j$, and hence $z_i \in (r_i, p_i)$. Now, if $x \in A$, there exists z_i, z_j which are in different components of $K - \{x\}$. Then, $x \in [z_i, z_j]$, which shows that $A \subseteq \bigcup_{i,j} [z_i, z_j]$; since Y is arcwise connected, $\bigcup_{i,j} [z_i, z_j] \subseteq Y$ and therefore $A \subseteq Y$; in particular, no point belonging to A is a terminal vertex of Y.

We want to see that Y contains exactly n terminal vertices and these are contained in the arcs (r_i, p_i). Let $y \in Y$ be a terminal vertex of Y. Since $y \notin L$, we have $y \neq p_i$ for all $1 \leq i \leq n$ and $y \notin A$ gives $y \in (r_i, p_i)$ for some i. For the above argument, it follows that

Y contains at most n terminal vertices; otherwise, two of them must belong to a same arc (r_i, p_i) which is not possible.

Now, given $1 \leq i \leq n$, $G_i = Y \cup [r_i, p_i]$ is a subspace of G. Since G is hereditary unicoherent, $Y \cap [r_i, p_i]$ is connected and non-degenerate (i.e., contains more than one point) because the arc $[r_i, z_i]$ is contained in the intersection and therefore such intersection is an arc whose extremes are r_i and say y_i. The point y_i is a terminal vertex of Y. This shows that Y contains at least n terminal vertices. We conclude $Y = C_y \in (\mathfrak{M}(A))^\circ$, where $y = (y_i)_{i=1}^n$.

Let us verify that n-cell $\mathfrak{M}(A)$ is actually maximal; for this purpose, suppose there exists a free n-cell $\mathcal{A} \subseteq C(G)$ such that $\mathfrak{M}(A) \subseteq \mathcal{A}$ with $\mathfrak{M}(A) \neq \mathcal{A}$. By Corollary 1, it must occur that there exists some point $C_x \in \partial \mathfrak{M}(A)$ such that $C_x \in \mathcal{A}^\circ$. Take $\varepsilon > 0$ such that $B(C_x, \varepsilon) \subseteq \mathcal{A}^\circ$.

Now, there are several cases to consider about the point C_x. The first one arises when we suppose $C_x = A \cup \left(\bigcup_{i=1}^n [r_i, x_i] \right)$, where, for some index, say $i = 1$, we have $x_1 = p_1$ is a terminal vertex of K and, at the same time, a ramification vertex of the dendrite G.

Let $u_i \in [r_i, x_i]$ (for $2 \leq i \leq n$) be points such that $[u_i, x_i] \subseteq B(x_i, \varepsilon)$ and let L_1, L_{n+1} be two different edges of G such that $L_1 \cap L_{n+1} \cap K = \{p_1\}$. Consider also points $u_1 \in L_1$ and $u_{n+1} \in L_{n+1}$ such that $[u_1, p_1] \subseteq B(p_1, \varepsilon)$ and $[u_{n+1}, p_1] \subseteq B(p_1, \varepsilon)$.

For $y_1 \in [p_1, u_1]$, $y_i \in [u_i, x_i]$ ($2 \leq i \leq n$) and $y_{n+1} \in [p_1, u_{n+1}]$, let $A_1 = [p_1, y_1]$, $A_i = [u_i, y_i]$ and $A_{n+1} = [p_i, y_{n+1}]$. The family of all subspaces of the form $A \cup \left(\bigcup_{i=1}^{n+1} A_i \right)$ is an $(n+1)$-cell contained in $B(C_x, \varepsilon) \subseteq \mathcal{A}^\circ$, and this is a contradiction. Similar considerations show that, if x_i is an essential point for some i, then it is possible find an $(n+1)$-cell contained in $\mathcal{A}^\circ$.

A second case is obtained when, for some index i, say $i = 1$, it occurs that $x_1 = r_1$. In this case, C_x can not belong to Int $(\mathfrak{M}(A))$ since this contradicts Lemma 5. However, if $C_x \in \text{Fr}(\mathfrak{M}(A))$, consider the decomposition $C(K) = \left(\bigcup_{T \in IT(K)} \mathfrak{M}(T) \right) \cup \left(\bigcup_{I \in E(K)} C(I) \right)$

of Lemma 7 (ii). We claim that we may suppose $C_x \in \left(\bigcup_{T \neq A} \mathfrak{M}(T) \right) \cup \left(\bigcup_{I \in E(K)} C(I) \right)$, where T runs over all internal trees of K different from A. Otherwise, there exists an open set $\mathcal{U}$ of $C(G)$ such that $C_x \in \mathcal{U} \subseteq C(G) - \left(\bigcup_{T \neq A} \mathfrak{M}(T) \right) \cup \left(\bigcup_{I \in E(K)} C(I) \right)$.

Let N be the first positive integer such that $B(C_x, \frac{1}{N}) \subseteq \mathcal{A}^\circ \cap \mathcal{U}$. Thus, for each $m \geq N$, there exists $Y_m \in C(G) - C(K)$, such that $H(Y_m, C_x) < \frac{1}{m}$. For each $m \geq N$, take a point $y_m \in Y_m - K$ and a point $x_m \in C_x$ such that $d(y_m, x_m) < \frac{1}{m}$. Since G is compact, the sequence (y_m) contains a convergent subsequence. We can suppose without loss of generality that (y_m) is actually convergent, say, to y. We claim that $y \in C_x$. Indeed, given $\varepsilon > 0$, choose $M \in \mathbb{N}$ such that $y_m \in B(y, \frac{\varepsilon}{2})$ for all $m \geq M$. If $m \geq M$ satisfies $\frac{1}{m} < \frac{\varepsilon}{2}$, then $d(x_m, y) \leq d(x_m, y_m) + d(y_m, y) < \frac{1}{m} + \frac{\varepsilon}{2} < \varepsilon$, that is, $x_m \in C_x \cap B(y, \varepsilon)$. With $\varepsilon > 0$ being arbitrary, we conclude that $y \in C_x$.

The above argument shows that y is a cluster point of $G - K$; this implies that $y = p_i$ for some index i, where $p_i \in R(G) \cup ES(G)$, and this case has already been analyzed.

Thus, we may suppose $C_x \in \left(\bigcup_{T \neq A} \mathfrak{M}(T) \right) \cup \left(\bigcup_{I \in E(K)} C(I) \right)$. In fact, by [10], [6.2, 6.3] or [12], [Lemma 2.6], we must suppose $C_x \in \left(\bigcup_{T \neq A} \partial \mathfrak{M}(T) \right) \cup \left(\bigcup_{I \in E(K)} \partial C(I) \right)$. Supposing first that $C_x \in \partial C(I)$ since the points belonging to the boundary manifold of a cell are cluster points of their interior manifold, we must have $\mathcal{A}° \cap (C(I))° \neq \emptyset$. Now, this set is open in $\mathcal{A}$, and, on the other hand, is contained in $C(I)$; this is impossible since $Dim(\mathcal{A}) = n > 2 = Dim(C(I))$.

Suppose now $C_x \in \partial \mathfrak{M}(T)$. Recall that A is the internal tree obtained from K by removing their terminal edges. It follows by [10], [5.3, 7.1] that $Dim(\mathfrak{M}(T)) < Dim(\mathfrak{M}(A))$. On the other hand, since C_x is a cluster point of $(\mathfrak{M}(T))°$, it must occur that $(\mathcal{A})° \cap (\mathfrak{M}(T))° \neq \emptyset$ and notice that this set is open in $C(G)$ and therefore open in $\mathcal{A}°$. Hence, there exists a homeomorphic copy of $[0,1]^n$ contained in $(\mathfrak{M}(T))°$, which is impossible regarding the dimension of $\mathfrak{M}(T)$.

The final case to consider is obtained when, for some index i, $x_i = p_i$ with $p_i \in T(G)$ or else $x_i \in (r_i, p_i)$. In this case, it is not difficult see that $C_x \in \text{Int}(\mathfrak{M}(T))$ contradicting Lemma 5. This shows that $\mathfrak{M}(A)$ is a maximal free n-cell as desired.

Conversely, let $\mathcal{A}$ be a free n-cell, $B \in \mathcal{A}°$ and let us analyze how B looks. Let $T(B) = \{p_1, \ldots, p_k\}$ and let $r_1, \ldots, r_s \in B - T(B)$ be the points such that $o_B(r_i) < o_G(r_i)$. Put $\alpha_i = o_G(r_i) - o_B(r_i)$ and assume that $k + \sum_{i=1}^{s} \alpha_i = m > n$. Consider $\varepsilon > 0$ such that $B(B, \varepsilon) \subseteq \mathcal{A}°$ and for each $1 \leq i \leq s$ consider also arcs $[u_{i_1}, r_i], \ldots, [u_{i_{\alpha_i}}, r_i]$ such that $[u_{i_j}, r_i] \subseteq B(r_i, \varepsilon)$ and $[u_{i_j}, r_i] \cap B = \{r_i\}$. In addition, take points v_t on the terminal edges of B such that $[v_t, p_t] \subseteq B(p_t, \varepsilon)$ for $1 \leq t \leq k$.

Letting $B_1 = B - \left(\bigcup_{i=1}^{n} [p_t, v_t] \right)$, we obtain that the family $\mathcal{H}$ of all subspaces of G has the form:

$$B_1 \cup \left(\bigcup_{t=1}^{k} [v_t, x_t] \right) \cap \left(\bigcup_{i=1}^{s} \left(\bigcup_{j=1}^{\alpha_i} [r_i, y_{i_j}] \right) \right),$$

where $x_t \in [v_t, p_t]$ and $y_{i_j} \in [r_i, u_{i_j}]$ is a m-cell contained in $\mathcal{A}$, which is absurd. Notice that the above argument in particular shows that $B - T(B)$ does not contain I-essential points. A similar reasoning shows that $B - T(B)$ also does not contain II-essential points. Now, assume that $m < n$. If $p_1, p_2, \ldots, p_q$ are the terminal vertices of B which are ordinal points of G, for each $1 \leq t \leq q$, let J_t be the edge of G such that $p_t \in J_t$ and for each $i \in \{1, \ldots, s\}$, let $I_{i_1}, I_{i_2}, \ldots, I_{i_{\alpha_i}}$ be the edges of G such that $J_{i_j} \cap B = \{r_i\}$. Hence, the tree

$$K = B \cup \left(\bigcup_{t=1}^{q} J_t \right) \cup \left(\bigcup_{i=1}^{s} \bigcup_{j=1}^{\alpha_i} I_{ij} \right)$$

has m terminal points and satisfies conditions (i) and (ii) and, by the *only if* part, we have already seen how to get a maximal free m-cell containing the above m-cell $\mathcal{H}$. Now, on the one hand, by Lemma 1, the cell $\mathcal{H}$ is free; on the other hand, since $\mathcal{H} \subseteq \mathcal{A}°$, Lemma 1 (a) gives that $\mathcal{H}$ is not a free cell and, again, this is absurd. Thus, we conclude that $m = n$ and K is the desired tree. □

5. Free Cells in Hyperspace of Dendroids

In this section, necessary and sufficient conditions are given so that Theorem 2 can be applied for dendroids. For this purpose, the notion of convergence space is required.

A non degenerated subspace A of a space X is called *convergence space* if there exists a sequence A_n of subspaces of X such that:

(1) $\lim A_n = A$,
(2) $A_n \cap A = \emptyset$.

The subspaces A_n can be chosen to be mutually disjoint (see [13] [5.11]).

Theorem 3. *Let G be a dendroid, a tree $K \subseteq G$, which satisfies the following conditions:*

(i) $T(K) = \{p_1, \ldots, p_n\} \subseteq R(G) \cup T(G) \cup ES(G)$,
(ii) *for all $x \in K - T(K)$, $o_K(x) = o_G(x)$.*

If A is the tree obtained from K by removing their terminal edges, then A induces a maximal free n-cell $\mathfrak{M}(A)$ if and only if this cell does not contain convergence subspaces.

Proof. The cell $\mathfrak{M}(A)$ is constructed as in the proof of Theorem 2. It is not hard to see that, if $C_x \in (\mathfrak{M}(A))°$ is a convergence subspace, then $\mathfrak{M}(A)$ can not be a free n-cell. On the other hand, if $\mathfrak{M}(A)$ is not a free n-cell, then there exists $Y = C_x \in (\mathfrak{M}(A))°$ such that, for each $\varepsilon > 0$, there exists $Z \in C(G) - (\mathfrak{M}(A))°$ with $H(Y, Z) < \varepsilon$.

Consider $\alpha_i = d(Y, p_i)$, $\beta = H(Y, \partial(\mathfrak{M}(A)))$, $\gamma_{T'} = H(Y, (\mathfrak{M}(A)))$ and $\delta_I = H(Y, C(I))$ (where T runs over the set of internal trees of K with $T \neq A$ and I runs over the set of edges of K). Since all these quantities are positive, take $\varepsilon > 0$ less than all of them and take $Z_1 \in C(X) - (\mathfrak{M}(A))$ such that $H(Z_1, Y) < \varepsilon$. If $Z_1 \cap Y \neq \emptyset$, we have the following cases:

(i) $Z_1 - K \neq \emptyset$,
in this case $p_i \in Z_1$ for some i. Hence, the ball $B(p_i, \varepsilon_1)$ intersects Y, and this contradicts the choice of ε_1.

(ii) $Z_1 \subseteq K$,
in this case, $Z_1 \in C(K) = \left[\bigcup_{T \in IT(K)} \mathfrak{M}(T) \right] \cup \left[\bigcup_{I \in E(K)} C(I) \right]$.

If $Z_1 \in \partial \mathfrak{M}(A)$, $Z_1 \in \mathfrak{M}(T)$ with $T \neq A$ or $Z_1 \in C(I)$, again this contradicts the choice of ε_1. Therefore, Z_1 and Y are disjoint. Taking $0 < \varepsilon_2 < H(Z_1, Y)$, in a similar way, we can obtain a subspace Z_2 with no points in common with Y and such that $H(Z_2, Y) < \varepsilon_2$. Continuing with this process, we obtain a sequence (Z_n) of mutually disjoint subspaces convergent to Y. □

6. Characterization of the Arc in Terms of Anchored Hyperspaces

The aim of this section is to prove that the arc is the unique arcwise connected space X, for which $C_p(X)$ is an arc for some $p \in X$ (Theorem 4). An important tool in the proof of this theorem is the use of order arcs. An *order arc* in 2^X is an arc α contained in 2^X such that, for any $A, B \in \alpha$, $A \subseteq B$ or $B \subseteq A$. The concepts and results we use for order arcs can be found in [13]. We use freely the notation found in there.

Proposition 1. *The anchored hyperspace $C_p(X)$ is an arc if and only if it is an order arc.*

Proof. Let α an order arc in $C(X)$ from $\{p\}$ to X. Since $p \in A$ for all $A \in \alpha$, we have $\alpha \subseteq C_p(CX)$. Now, it is sufficient to show that $\{p\}$ and X are also the end points of $C_p(X)$, and this will be done by proving that neither $\{p\}$ nor X are cut points of $C_p(X)$ (see [16], [Theorem 1, Pag. 179]). Take different points $A, B \in C_p(X) - \{p\}$ if β and γ are order arcs from A to $A \cup B$ and from B to $A \cup B$ respectively, then $\beta \cup \gamma \subseteq C_p(X) - \{p\}$ is an arc containing the points A and B; this shows that $\{p\}$ is not a cut point of $C_p(X)$. Similarly, if $A, B \in C_p(X) - \{X\}$, taking β and γ order arcs from $\{p\}$ to A and from $\{p\}$ to B, one obtains that X is not a cut point either and therefore $\alpha = C_p(X)$. The converse is obvious. □

A point p of a space X is an *irreducibility point* of X if there exists another point q such that no proper subspace contains both points. The following result is due to Kuratoski and is a handy tool in the proof of Theorem 4.

Lemma 8 (Kuratoski's Theorem, [15])**.** *Let X be a space and let $p \in X$. Then, p is point of irreducibility of X if and only if X is not the union of two proper subspaces both of which contain p.*

Theorem 4. *Let X be an arcwise connected space. Then, $C_p(X)$ is an arc for some $p \in X$ if and only if X is an arc.*

Proof. By Proposition 1, $C_p(X)$ is an order arc from $\{p\}$ to X. It follows that X is not the union of two proper subspaces both containing the point p. By Lemma 8, it turns out that

p is an irreducibility point of X, if $q \in X$ is another point such that no proper subspace of X contains the points p and q; the arcwise connectedness implies that $X = [p, q]$.

For the converse, suppose without loss of generality that $X = [0, 1]$. Letting $p = 0$, the map
$$x \mapsto [p, x],$$
is a homeomorphism from X to $C_p(X)$. □

The arcwise connectedness hypothesis is necessary in the above theorem (see [13] [Example 1.1]).

7. Comparative Studies and Conclusions

Some of the main goals on hyperspace research from a theoretical approach are: to obtain topological models corresponding to familiar or not difficult to handle spaces, to find relations between hyperspaces and their underlying spaces, uniqueness of hyperspaces, i.e., to investigate which spaces are the only ones whose hyperspaces possess a given structure. Motivated by the studies carried out in [17,18], the present work was deemed convenient by the authors. In the aforementioned works, the existence of cells in hyperspaces is characterized. Our work is carried out on infinite graphs and describes when such cells are free.

In [19], the arc is characterized in terms of anchored hyperspaces within the class of trees. In our work, we conduct a similar study but within a broader class of spaces, the arc-connected spaces.

Question: If the class of anchored hyperspaces of an arcwise connected space X matches the class of anchored hyperspaces of a connected graph G, does it follow that X and G are homeomorphic?

Author Contributions: Writing—original draft, J.Á.J.M., G.R.H., J.R.V. and O.R.C. All authors contributed equally to this work. All authors have read and agreed to the published version of the manuscript.

Funding: This article was supported by the Perfil Deseable del Programa para el Desarrollo Profesional Docente (PRODEP), a federal institution of México's government and the Universidad Autónoma de Guerrero (UAGro).

Institutional Review Board Statement: Not applicable.

Informed Consent Statement: Not applicable.

Data Availability Statement: Not applicable.

Conflicts of Interest: The authors declare no conflict of interest.

References

1. Charatonik, W. J.; Fernández-Bayort, T.; Quintero, A. Hyperspaces of generalized continua which are infinite cylinders. *Topol. Appl.* **2019**, *267*, 106829. [CrossRef]
2. Eberhart, C. Continua with locally connected Whitney continua. *Houst. J. Math.* **2019**, *4*, 165–173.
3. Bryant, D.; Horsley, D.; Maenhaut, B.; Smith, B.R. Descompositions of complete multigraphs into cycles of varying lenght. *J. Comb. Theory Ser. B* **2018**, *129*, 79–106. [CrossRef]
4. Gentner, M.; Rautenbach, D. Feedback vertex sets in cubic multigraphs. *Discret. Math.* **2015**, *338*, 2179–2185. [CrossRef]
5. Hernández, J.C.; Reyna, G.; Romero, J.; Rosario, O. Transitivity on minimum dominating sets of paths and cycles. *Symmetry* **2020**, *12*, 2053. [CrossRef]
6. Klee, S.; Nevo, E.; Novik, I.; Zheng, H. A lower bound theorem for centrally symmetric simplicial polytopes. *Discret. Comput. Geom.* **2018**, *61*, 541–561. [CrossRef]
7. Ramírez, A.; Reyna, G.; Rosario, O. Spectral study of the inverse index. *Adv. Appl. Discret. Math.* **2018**, *3*, 195–211. [CrossRef]
8. Gallo, G.; Longo, G.; Pallottino, S.; Nguyen, S. Directed hypergraphs and applications. *Discret. Appl. Math.* **1993**, *42*, 177–201. [CrossRef]
9. Feng, K. Spectra of hypergraphs and applications. *J. Number Theory* **1996**, *60*, 1–22. [CrossRef]
10. Duda, R. On the hyperspace of subcontinua of a finite graph I. *Fundam. Math.* **1968**, *62*, 265–286. [CrossRef]
11. Eberhart, C. Intervals of continua which are Hilbert cubes. *Proc. Am. Math. Soc.* **1978**, *68*, 220–224. [CrossRef]

12. Reyna, G.; Romero, J.; Espinobarros, I. Anchored hyperspaces and multigraphs. *Contrib. Discret. Math.* **2019**, *1*, 150–166.
13. Nadler, S.B., Jr. *Hyperspaces of Sets: A Text with Research Questions*, 1st ed.; Marcel Dekker Inc.: New York, NY, USA, 1978.
14. Bermudo, S.; Rodríguez, J.M.; Sigarreta, J.M.; Vilaire, J.M. Gromov hyperbolic graphs. *Discret. Math.* **2013**, *313*, 1575–1585. [CrossRef]
15. Nadler, S.B. *Continuum Theory. An Introduction*, 1st ed.; CRC Press: New York, NY, USA, 1992.
16. Kuratowski, K. *Topology II*; Academic Press: New York, NY, USA, 1966.
17. Rogers, J.T. Dimension of hyperspace. *Bull. Pol. Acad. Sci.* **1972**, *20*, 177–179.
18. Illanes, A. Cells and cubes in hyperspaces. *Fundam. Math.* **1988**, *130*, 57–65. [CrossRef]
19. Corona–Vázquez, F.; Quiñones–Estrella, R.A.; Sánchez-Martínez, J.; Toalá-Enríquez, R. Uniqueness of the hyperspaces C (p, X) in the class of trees. *Topol. Appl.* **2020**, *269*, 106926. [CrossRef]

Article

Local Inclusive Distance Vertex Irregular Graphs

Kiki Ariyanti Sugeng [1,*,†], Denny Riama Silaban [1,†], Martin Bača [2,†] and Andrea Semaničová-Feňovčíková [2,†]

1. Department of Mathematics, Faculty of Mathematics and Natural Sciences, Universitas Indonesia, Kampus UI Depok, Depok 16424, Indonesia; denny@sci.ui.ac.id
2. Department of Applied Mathematics and Informatics, Technical University, 042 00 Košice, Slovakia; martin.baca@tuke.sk (M.B.); andrea.fenovcikova@tuke.sk (A.S.-F.)
* Correspondence: kiki@sci.ui.ac.id
† These authors contributed equally to this work.

Abstract: Let $G = (V, E)$ be a simple graph. A vertex labeling $f : V(G) \to \{1, 2, \ldots, k\}$ is defined to be a local inclusive (respectively, non-inclusive) d-distance vertex irregular labeling of a graph G if for any two adjacent vertices $x, y \in V(G)$ their weights are distinct, where the weight of a vertex $x \in V(G)$ is the sum of all labels of vertices whose distance from x is at most d (respectively, at most d but at least 1). The minimum k for which there exists a local inclusive (respectively, non-inclusive) d-distance vertex irregular labeling of G is called the local inclusive (respectively, non-inclusive) d-distance vertex irregularity strength of G. In this paper, we present several basic results on the local inclusive d-distance vertex irregularity strength for $d = 1$ and determine the precise values of the corresponding graph invariant for certain families of graphs.

Keywords: (inclusive) distance vertex irregular labeling; local (inclusive) distance vertex irregular labeling

MSC: 05C15; 05C78

1. Introduction

All graphs considered in this paper are simple finite. We use $V(G)$ for the vertex set and $E(G)$ for the edge set of a graph G. The neighborhood $N_G(x)$ of a vertex $x \in V(G)$ is the set of all vertices adjacent to x, which is a set of vertices whose distance from x is 1. Otherwise, $N_G[x]$ denotes the set of all neighbors of a vertex $x \in V(G)$ including x, which is the set of vertices whose distance from x is at most 1. We are following the standard notation and the terminology presented in [1].

The notion of the irregularity strength was introduced by Chartrand et al. in [2]. For a given edge k-labeling $\alpha : E(G) \to \{1, 2, \ldots, k\}$, where k is a positive integer, the associated weight of a vertex $x \in V(G)$ is $w_\alpha(x) = \sum_{y \in N_G(x)} \alpha(xy)$. Such a labeling α is called *irregular* if $w_\alpha(x) \neq w_\alpha(y)$ for every pair x, y of vertices of G. The smallest integer k for which an irregular labeling of G exists is known as the *irregularity strength* of G. This parameter has attracted much attention, see [3–5].

Inspired by irregularity strength and distance magic labeling defined in [6] and investigated in [7], Slamin [8] introduced the concept of a distance vertex irregular labeling of graphs. A *distance vertex irregular labeling* of a graph is a mapping $\beta : V(G) \to \{1, 2, \ldots, k\}$ such that the set of vertex weights consists of distinct numbers, where the weight of a vertex $x \in V(G)$ under the labeling β is defined as $wt_\beta(x) = \sum_{y \in N_G(x)} \beta(y)$. The minimum k for which a graph G has a distance vertex irregular labeling is called the *distance vertex irregularity strength* of G and is denoted by $\mathrm{dis}(G)$.

In [8], Slamin determined the exact value of the distance vertex irregularity strength for complete graphs, paths, cycles and wheels, namely $\mathrm{dis}(K_n) = n$, for $n \geq 3$, $\mathrm{dis}(P_n) = \lceil n/2 \rceil$, for $n \geq 4$, $\mathrm{dis}(C_n) = \lceil (n+1)/2 \rceil$, for $n \equiv 0, 1, 2, 3 \pmod{8}$ and $\mathrm{dis}(W_n) =$

$\lceil(n+1)/2\rceil$, for $n \equiv 0, 1, 2, 5 \pmod 8$. Completed results for cycles and wheels are proved in [9].

Bong et al. [10] generalized the concept of a distance vertex irregular labeling to inclusive and non-inclusive d-distance vertex irregular labelings. The difference between inclusive and non-inclusive labeling depends on the way whether the vertex label is included in the vertex weight or not. The symbol d represents how far the neighborhood is considered. Thus, an inclusive (respectively, non-inclusive) d-distance vertex irregular labeling of a graph G is a mapping β such that the set of vertex weights consists of distinct numbers, where the weight of a vertex $x \in V(G)$ is the sum of all labels of vertices whose distance from x is at most d (respectively, at most d but at least 1). The minimum k for which there exists an inclusive (respectively, non-inclusive) d-distance vertex irregular labeling of a graph G is called the *inclusive (respectively, non-inclusive) d-distance vertex irregularity strength* of G. The non-inclusive 1-distance vertex irregularity strength of a graph G is using Slamin's [8] terminology known as the distance vertex irregularity strength of G, denoted by $\mathrm{dis}(G)$. For the inclusive 1-distance vertex irregularity strength, we will use notation $\mathrm{idis}(G)$.

In [10] is determined the inclusive 1-distance vertex irregularity strength for paths P_n, $n \equiv 0 \pmod 3$, stars, double stars $S(m,n)$ with $m \leq n$, a lower bound for caterpillars, cycles, and wheels. In [11] is established a lower bound of the inclusive 1-distance vertex irregularity strength for any graph and determined the exact value of this parameter for several families of graphs, namely for complete and complete bipartite graphs, paths, cycles, fans, and wheels. More results on triangular ladder and path for $d \geq 1$ has been proved in [12,13].

Motivated by a distance vertex labeling [8], an irregular labeling [2] and a recent paper on a local antimagic labeling [14], we introduce in this paper the concept of local inclusive and local non-inclusive d-distance vertex irregular labelings.

A vertex labeling $f : V(G) \to \{1, 2, \ldots, k\}$ is defined to be a local *inclusive* (respectively, *non-inclusive*) *d-distance vertex irregular labeling* of a graph G if for any two adjacent vertices $x, y \in V(G)$ their weights are distinct, where the weight of a vertex $x \in V(G)$ is the sum of all labels of vertices whose distance from x is at most d (respectively, at most d but at least 1). The minimum k for which there exists a local inclusive (respectively, non-inclusive) d-distance vertex irregular labeling of G is called the *local inclusive* (respectively, *non-inclusive*) *d-distance vertex irregularity strength* of G and denoted by $\mathrm{lidis}_d(G)$ (respectively, $\mathrm{ldis}_d(G)$). If there is no such labeling for the graph G then the value of $\mathrm{lidis}_d(G)$ is defined as ∞. In the case when $d = 1$ the index d can be omitted, thus $\mathrm{lidis}_1(G) = \mathrm{lidis}(G)$ (respectively, $\mathrm{ldis}_1(G) = \mathrm{ldis}(G)$). In this paper, we only discuss the case for inclusive labeling with $d = 1$. Note that the concept of a local non-inclusive distance vertex irregular labeling has been introduced earlier in [15] with a different name. For more information about labeled graphs see [16].

In this paper, we present several basic results and some estimations on the local inclusive 1-distance vertex irregularity strength and determine the precise values of the corresponding graph invariant for several families of graphs.

2. Basic Properties

In the following observations, we give several basic properties of $\mathrm{lidis}(G)$. The first observation gives a relation between the local inclusive distance vertex irregularity strength, $\mathrm{lidis}(G)$, and the inclusive distance vertex irregularity strength, $\mathrm{idis}(G)$. The second and third observations give the requirement for giving the label of two vertices which have a common neighbor.

Observation 1. *For a graph G, it holds that* $\mathrm{lidis}(G) \leq \mathrm{idis}(G)$.

Observation 2. *If there exists an edge uv in a graph G such that $N_G(u) - \{v\} = N_G(v) - \{u\}$, then for any local non-inclusive distance vertex irregular labeling f of a graph G holds $f(u) \neq f(v)$.*

Observation 3. *If there exists an edge uv in a graph G such that $N_G(u) - \{v\} = N_G(v) - \{u\}$, then $\text{lidis}(G) = \infty$.*

The next theorem gives a sufficient and necessary condition for $\text{lidis}(G) < \infty$. Note that the graph G is not necessarily connected.

Theorem 1. *For a graph G, it holds that $\text{lidis}(G) = \infty$ if and only if there exists an edge $uv \in E(G)$ such that $N_G[u] = N_G[v]$.*

Proof. If there exists an edge $uv \in E(G)$ such that $N_G[u] = N_G[v]$, then immediately follows Observation 3 and we obtain $\text{lidis}(G) = \infty$. On the other hand, if $\text{lidis}(G) = \infty$ then there exist at least two vertices u and v in G that have the same weight under any vertex labeling. It is only happened if $N_G[u] = N_G[v]$. □

Immediately from the previous theorem we obtain the following result.

Corollary 1. *If there exist two distinct vertices u, v in G such that $\deg_G(u) = \deg_G(v) = |V(G)| - 1$, then $\text{lidis}(G) = \infty$.*

Thus, for complete graphs we obtain

Corollary 2. *Let n be a positive integer. Then*

$$\text{lidis}(K_n) = \begin{cases} 1, & \text{if } n = 1, \\ \infty, & \text{if } n \geq 2. \end{cases}$$

Now, we present a sufficient and necessary condition for $\text{lidis}(G) = 1$.

Theorem 2. *Let G be a graph. Then $\text{lidis}(G) = 1$ if and only if for every edge $uv \in E(G)$, $\deg(u) \neq \deg(v)$.*

Proof. Consider a labeling that assigns number 1 to every vertex of a graph G. Under this labeling, the weight of any vertex v in G is $wt(v) = \deg_G(v) + 1$. Thus, adjacent vertices can have distinct weights if and only if they have distinct degrees. □

The chromatic number of a graph G, denoted by $\chi(G)$, is the smallest number of colors needed to color the vertices of G so that no two adjacent vertices share the same color, see [1]. The following result gives a trivial lower bound for the number of distinct induced vertex weights under any local inclusive distance vertex irregular labeling of a graph G.

Theorem 3. *For a graph G, the number of distinct induced vertex weights under any local inclusive distance vertex irregular labeling is at least $\chi(G)$.*

3. Local Inclusive Distance Vertex Irregularity Strength for Several Families of Graphs

In this section, we provide the exact values of local inclusive distance vertex irregularity strengths of some standard graphs such as paths, cycles, complete bipartite graphs, complete multipartite graphs, and caterpillars. We also give results on several products of graphs, such as corona graphs, union graphs, and join product graphs.

Theorem 4. *Let C_n be a cycle on n vertices $n \geq 3$. Then*

$$\text{lidis}(C_n) = \begin{cases} \infty, & \text{if } n = 3, \\ 2, & \text{if } n \text{ is even}, \\ 3, & \text{if } n \text{ is odd}, n \geq 5. \end{cases}$$

Proof. Let $V(C_n) = \{v_i : i = 1, 2, \ldots, n\}$ be the vertex set and let $E(C_n) = \{v_i v_{i+1} : i = 1, 2, \ldots, n-1\} \cup \{v_1 v_n\}$ be the edge set of a cycle C_n. The lower bound for the local inclusive distance vertex irregularity strength of C_n follows from Theorem 3 as

$$\chi(C_n) = \begin{cases} 3, & \text{if } n \text{ is odd,} \\ 2, & \text{if } n \text{ is even.} \end{cases}$$

As C_3 is isomorphic to K_3 we use Corollary 2 in this case.
For n even, we label the vertices of C_n as follows

$$f(v_i) = \begin{cases} 1, & \text{if } i \text{ is odd,} \\ 2, & \text{if } i \text{ is even.} \end{cases}$$

Then, for the vertex weights we obtain

$$wt_f(v_i) = \begin{cases} 5, & \text{if } i \text{ is odd,} \\ 4, & \text{if } i \text{ is even.} \end{cases}$$

Thus, for n even we obtain $\text{lidis}(C_n) = 2$.
For $n = 5$, we label the vertices such that $f(v_1) = f(v_2) = 1$, $f(v_3) = 3$ and $f(v_4) = f(v_5) = 2$. Then, $wt_f(v_1) = 4$, $wt_f(v_2) = wt_f(v_5) = 5$, $wt_f(v_3) = 6$ and $wt_f(v_4) = 7$. Thus, $\text{lidis}(C_5) = 3$.
For n odd, $n \geq 7$, the vertices are labeled in the following way

$$f(v_i) = \begin{cases} 1, & \text{if } i \text{ is odd, } 1 \leq i \leq n-4, \\ 2, & \text{if } i \text{ is even, } 2 \leq i \leq n-3, \\ 3, & \text{if } i = n-2, n-1, n. \end{cases}$$

The weights of vertices are

$$wt_f(v_i) = \begin{cases} 6, & \text{if } i = 1, n-3, \\ 5, & \text{if } i \text{ is odd, } 3 \leq i \leq n-4, \\ 4, & \text{if } i \text{ is even, } 2 \leq i \leq n-5, \\ 8, & \text{if } i = n-2, \\ 9, & \text{if } i = n-1, \\ 7, & \text{if } i = n. \end{cases}$$

As adjacent vertices have distinct weights we obtain $\text{lidis}(C_n) = 3$ for n odd. The above explanation concludes the proof. □

Corollary 3. *Let P_n be a path on n vertices $n \geq 2$. Then*

$$\text{lidis}(P_n) = \begin{cases} \infty, & \text{if } n = 2, \\ 2, & \text{if } n \geq 3. \end{cases}$$

Proof. Let $V(P_n) = \{v_i : i = 1, 2, \ldots, n\}$ be the vertex set and let $E(P_n) = \{v_i v_{i+1} : i = 1, 2, \ldots, n-1\}$ be the edge set of a path P_n. The result for $n = 2$ follows from Corollary 2.
For $n \geq 3$, according to Theorem 3, the $\text{lidis}(P_n)$ should be more than one. Using the vertex labels for n even as in Theorem 4 and the corresponding vertex weights are

$$wt_f(v_i) = \begin{cases} 3, & \text{if } i = 1, n, \\ 4, & \text{if } i \text{ is even, } i \neq n, \\ 5, & \text{if } i \text{ is odd, } i \neq 1 \text{ and } i \neq n. \end{cases}$$

Thus, lidis$(P_n) = 2$. □

The following result deals with complete multipartite graphs.

Theorem 5. *Let $K_{n_1,n_2,\ldots,n_m}$ be a complete multipartite graph, $n_i \geq 1, i = 1, 2, \ldots, m, m \geq 2$. Then,*

$$\text{lidis}(K_{n_1,n_2,\ldots,n_m}) = \begin{cases} \infty, & \text{if } 1 = n_1 = n_2, \\ 1, & \text{if } n_1 < n_2 < \cdots < n_m, \\ m, & \text{if } 2 \leq n_1 = n_2 = \cdots = n_m. \end{cases}$$

Proof. Let us denote the vertices in the independent set $V_i, i = 1, 2, \ldots, m$ of a complete multipartite graph $K_{n_1,n_2,\ldots,n_m}$ by symbols $v_i^1, v_i^2, \ldots, v_i^{n_i}$.

If $1 = n_1 = n_2$, then the vertices v_1^1 and v_2^1 have the same degrees

$$\deg(v_1^1) = \deg(v_2^1) = \sum_{j=3}^{m} n_j + 1 = |V(K_{n_1,n_2,\ldots,n_m})| - 1$$

and thus, by Corollary 1 we obtain lidis$(K_{n_1,n_2,\ldots,n_m}) = \infty$.

If $n_1 < n_2 < \cdots < n_m$, then all adjacent vertices have distinct degrees. More precisely, the degree of a vertex $v_i^j, i = 1, 2, \ldots, m, j = 1, 2, \ldots, n_i$ is $\deg(v_i^j) = \sum_{j=1}^{m} n_j - n_i + 1$. Thus, by Theorem 2, we obtain lidis$(K_{n_1,n_2,\ldots,n_m}) = 1$.

If $2 \leq n_1 = n_2 = \cdots = n_m = n$ consider a vertex labeling f of $K_{n_1,n_2,\ldots,n_m}$ defined such that

$$f(v_i^j) = i$$

for $i = 1, 2, \ldots, m, j = 1, 2, \ldots, n$ and the corresponding vertex weights are

$$wt_f(v_i^j) = \frac{nm(m+1)}{2} - (n-1)i.$$

Thus, all adjacent vertices have distinct weights. Thus, lidis$(K_{n_1,n_2,\ldots,n_m}) \leq m$. Using mathematical induction, it is not complicated to show that lidis$(K_{n_1,n_2,\ldots,n_m}) \geq m$. This concludes the proof. □

The following corollary gives the exact value of the studied parameter for complete bipartite graphs.

Corollary 4. *Let $K_{m,n}, 1 \leq m \leq n$, be a complete bipartite graph. Then*

$$\text{lidis}(K_{m,n}) = \begin{cases} \infty, & \text{if } m = n = 1, \\ 2, & \text{if } m = n \geq 2, \\ 1, & \text{if } m \neq n. \end{cases}$$

The *corona product* of G and H is the graph $G \odot H$ obtained by taking one copy of G, called the center graph along with $|V(G)|$ copies of H, called the outer graph, and making the ith vertex of G adjacent to every vertex of the ith copy of H, where $1 \leq i \leq |V(G)|$. For arbitrary graphs G, we can prove the following result.

Theorem 6. *Let r be a positive integer. Then, for $r \geq 2$ holds*

$$\text{lidis}(G \odot \overline{K_r}) \leq \text{lidis}(G).$$

Moreover, if G is a graph with no component of order 1 then also lidis$(G \odot K_1) \leq$ lidis(G).

Proof. If $\text{lidis}(G) = \infty$ then by Theorem 1 there exists at least one edge $uv \in E(G)$ such that $N_G[u] = N_G[v]$. However, as for $r \geq 2$ or for $r = 1$ if G has no component of order 1, in $G \odot \overline{K_r}$ all vertices have distinct closed neighborhood and thus $\text{lidis}(G \odot \overline{K_r}) < \infty$.

Now, consider that $\text{lidis}(G) < \infty$ and let f be a local inclusive distance vertex irregular labeling of G. We define a labeling g of $G \odot \overline{K_r}$ such that

$$g(v) = f(v), \quad \text{if } v \in V(G),$$
$$g(v) = 1, \quad \text{if } \deg_{G \odot \overline{K_r}}(v) = 1.$$

For the vertex weights, we obtain

$$wt_g(v) = wt_f(v) + r, \quad \text{if } v \in V(G),$$
$$wt_g(v) = 1 + f(u), \quad \text{if } \deg_{G \odot \overline{K_r}}(v) = 1 \text{ and } uv \in E(G \odot \overline{K_r}).$$

Evidently, for $r \geq 2$ or for $r = 1$ if G has no component of order 1, i.e., $\deg_G(v) \geq 1$ for every $v \in V(G)$, we obtain that under the labeling g the vertex weights of adjacent vertices are different. □

Moreover, we can prove that the parameter $\text{lidis}(G \odot \overline{K_r})$ is finite except the case when $G \odot \overline{K_r}$ contains a component isomorphic to K_2.

Theorem 7. *Let r be a positive integer. Then,*

$$\text{lidis}(G \odot \overline{K_r}) \leq |V(G)|$$

except the case when $r = 1$ and the graph G contains a component of order 1.

Proof. Let us denote the vertices of a graph G by symbols $v_1, v_2, \ldots, v_{|V(G)|}$ such that for every $i = 1, 2, \ldots, |V(G)| - 1$ holds

$$\deg_G(v_i) \leq \deg_G(v_{i+1})$$

and let v_i^j, $j = 1, 2, \ldots, r$ be the vertices of degree 1 adjacent to v_i, $i = 1, 2, \ldots, |V(G)|$, in $G \odot \overline{K_r}$. Now, we define a labeling f that assigns 1 to every vertex of G. Thus, for every $i = 1, 2, \ldots, |V(G)|$

$$wt_f(v_i) = \deg_G(v_i) + 1.$$

We extend the labeling f of the graph G to the labeling g of the graph $G \odot \overline{K_r}$ in the following way

$$g(v_i) = f(v_i), \quad \text{if } i = 1, 2, \ldots, |V(G)|,$$
$$g(v_i^j) = i, \quad \text{if } i = 1, 2, \ldots, |V(G)|, j = 1, 2, \ldots, r.$$

The induced vertex weights are

$$wt_g(v_i) = \deg_G(v_i) + 1 + ri, \quad \text{if } i = 1, 2, \ldots, |V(G)|,$$
$$wt_g(v_i^j) = 1 + i, \quad \text{if } i = 1, 2, \ldots, |V(G)|, j = 1, 2, \ldots, r.$$

For $r \geq 2$ and for $r = 1$ if the graph G has no component of order 1, i.e., $\deg(v_i) \geq 1$ for every $i = 1, 2, \ldots, |V(G)|$, we obtain that all adjacent vertices have distinct weights. □

Note that the upper bound in the previous theorem is tight, since $\text{lidis}(K_n \odot K_1) = n$. Immediately, from Theorem 2, we have the following result

Theorem 8. *For $r \geq 2$ it holds $\text{lidis}(G \odot \overline{K_r}) = 1$ if and only if $\text{lidis}(G) = 1$.*

Moreover, when G has no component of order 1 then $\text{lidis}(G \odot \overline{K_1}) = 1$ if and only if $\text{lidis}(G) = 1$.

Now, we present results for corona product of paths, cycles, and complete graphs with totally disconnected graph $\overline{K_r}$, $r \geq 1$. Combining Theorems 3 and 6, we obtain

Theorem 9. *Let P_n be a path on n vertices $n \geq 2$ and let r be a positive integer. Then*

$$\text{lidis}(P_n \odot \overline{K_r}) = 2.$$

Theorem 10. *Let C_n be a cycle on n vertices $n \geq 3$ and let r be a positive integer. Then*

$$\text{lidis}(C_n \odot \overline{K_r}) = \begin{cases} 3, & \text{if } n = 3 \text{ and } r = 1, \\ 2, & \text{otherwise.} \end{cases}$$

Proof. Let

$$V(C_n \odot \overline{K_r}) = \{v_i : i = 1, 2, \ldots, n\} \cup \{v_i^j : i = 1, 2, \ldots, n; j = 1, 2, \ldots, r\}$$

be the vertex set and let

$$E(C_n \odot \overline{K_r}) = \{v_i v_{i+1} : i = 1, 2, \ldots, n-1\} \cup \{v_1 v_n\}$$
$$\cup \{v_i v_i^j : i = 1, 2, \ldots, n; j = 1, 2, \ldots, r\}$$

be the edge set of $C_n \odot \overline{K_r}$.

For even n the result follows from Theorems 4 and 6. For $n = 3$ and $r = 1$ consider the labeling illustrated on Figure 1.

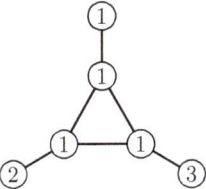

Figure 1. A local inclusive distance vertex irregular labeling of $C_3 \odot \overline{K_1}$.

For odd n and $(n, r) \neq (3, 1)$, we define a vertex labeling f of $C_n \odot \overline{K_r}$ such that

$$f(v_i) = \begin{cases} 2, & \text{for } i = 1, \\ 1, & \text{for } i = 2, 3, \ldots, n, \end{cases}$$

$$f(v_i^j) = \begin{cases} 2, & \text{for } i = 2, 4, \ldots, n-1, n \text{ and } j = 1, \\ 1, & \text{otherwise.} \end{cases}$$

The weights of vertices of degree $r + 2$ are

$$wt_f(v_i) = \begin{cases} r+3, & \text{for } i = 3, 5, \ldots, n-2, \\ r+4, & \text{for } i = 1, 4, 6, \ldots, n-1, \\ r+5, & \text{for } i = 2, n. \end{cases}$$

As the weights of vertices of degree one are either 2 or 3, we obtain that adjacent vertices have distinct weights. □

Theorem 11. Let n, r be positive integers. Then

$$\text{lidis}(K_n \odot \overline{K_r}) = \begin{cases} \infty, & \text{if } n = 1, r = 1, \\ 1 + \left\lceil \frac{n-1}{r} \right\rceil, & \text{otherwise.} \end{cases}$$

Proof. As the graph $K_1 \odot \overline{K_1}$ is isomorphic to the complete graph K_2 we use Corollary 2 in this case.

Let $(n, r) \neq (1, 1)$. Let the vertex set and the edge set of $K_n \odot \overline{K_r}$ be the following

$$V(K_n \odot \overline{K_r}) = \{v_i, v_i^j : i = 1, 2, \ldots, n; j = 1, 2, \ldots, r\},$$
$$E(K_n \odot \overline{K_r}) = \{v_i v_j : i = 1, 2, \ldots, n-1; j = i+1, i+2, \ldots, n\}$$
$$\cup \{v_i v_i^j : i = 1, 2, \ldots, n; j = 1, 2, \ldots, r\}.$$

We define a vertex labeling f of $K_n \odot \overline{K_r}$ such that

$$f(v_i) = 1 + \left\lceil \frac{n-1}{r} \right\rceil, \quad \text{if } i = 1, 2, \ldots, n,$$

$$f(v_i^j) = \begin{cases} 1 + \left\lceil \frac{i-1}{r} \right\rceil, & \text{if } i = 1, 2, \ldots, n, j = 1, 2, \ldots, A_i, \\ 1 + \left\lfloor \frac{i-1}{r} \right\rfloor, & \text{if } i = 1, 2, \ldots, n, j = A_i + 1, A_i + 2, \ldots, r, \end{cases}$$

where for every $i = 1, 2, \ldots, n$ the parameter A_i, $1 \leq A_i \leq r$, is defined such that

$$i - 1 \equiv A_i \pmod{r}.$$

For the vertex weights, we obtain

$$wt_f(v_i) = n\left(1 + \left\lceil \frac{n-1}{r} \right\rceil\right) + r + i - 1, \quad \text{if } i = 1, 2, \ldots, n,$$

$$wt_f(v_i^j) = \begin{cases} \left\lceil \frac{n-1}{r} \right\rceil + 2 + \left\lceil \frac{i-1}{r} \right\rceil, & \text{if } i = 1, 2, \ldots, n, j = 1, 2, \ldots, A_i, \\ \left\lceil \frac{n-1}{r} \right\rceil + 2 + \left\lfloor \frac{i-1}{r} \right\rfloor, & \text{if } i = 1, 2, \ldots, n, j = A_i + 1, A_i + 2, \ldots, r. \end{cases}$$

Evidently adjacent vertices have distinct weights. Thus, as the maximal vertex label is $1 + \lceil (n-1)/r \rceil$, the proof is completed. □

A *caterpillar* is a graph derived from a path by hanging any number of leaves from the vertices of the path. We denote the caterpillar as $S_{n_1, n_2, \ldots, n_r}$, where the vertex set is $V(S_{n_1, n_2, \ldots, n_r}) = \{c_i : 1 \leq i \leq r\} \cup \bigcup_{i=1}^{r} \{u_i^j : 1 \leq j \leq n_i\}$, and the edge set is $E(S_{n_1, n_2, \ldots, n_r}) = \{c_i c_{i+1} : 1 \leq i \leq r-1\} \cup \bigcup_{i=1}^{r} \{c_i u_i^j : 1 \leq j \leq n_i\}$.

Theorem 12. For every caterpillar $S_{n_1, n_2, \ldots, n_r}$ with at least 3 vertices holds $\text{lidis}(S_{n_1, n_2, \ldots, n_r}) \leq 2$.

Proof. For a regular caterpillar, thus the case $n_1 = n_2 = \ldots = n_r = n$, using Theorem 9, we obtain that $\text{lidis}(S_{n,n,\ldots,n}) = 2$.

For the other cases, label the vertices of a caterpillar $S_{n_1, n_2, \ldots, n_r}$ using the following algorithm.

Step 1: Label all vertices with 1.
Then the weights of vertices c_i, $i = 1, 2, \ldots, r$ are $\deg(c_i)$ and all vertices of degree 1 have weight 2.
Step 2: Find the smallest index s, $2 \leq s \leq r - 1$, such that $wt(c_{s+1}) = wt(c_s)$.
Step 3: If such number does not exist, it means that adjacent vertices have distinct degrees and thus $\text{lidis}(S_{n_1, n_2, \ldots, n_r}) = 1$. We are done.

Step 4: If such number exists either relabel a leaf of adjacent to c_{s+1} (if a leaf exists) from 1 to 2 or relabel the vertex c_{s+2} from 1 to 2. Then $wt(c_{s+1}) = wt(c_s) + 1$.
Note that this relabeling will not have an effect on weights of vertices c_i for every $i \leq s$.
Step 5: Find the smallest index t, $s + 1 \leq t \leq r - 1$, such that $wt(c_{t+1}) = wt(c_t)$.
Step 6: If such number does not exist, it means that adjacent vertices have distinct degrees and thus $\text{lidis}(S_{n_1,n_2,\ldots,n_r}) = 2$. We are finished.
Step 7: If such number exists either relabel a leaf of adjacent to c_{t+1} (if a leaf exists) from 1 to 2 or relabel the vertex c_{t+2} from 1 to 2. Then $wt(c_{s+1}) = wt(c_t) + 1$.
Step 8: Return to Step 5.

After a finite number of steps, the algorithm stops and the weights of the vertices are always different from the weights of their neighbors. □

A similar algorithm can be used to obtain a result for closed caterpillars, which are graphs where the removal of all pendant vertices gives a cycle. We denote the closed caterpillar as $CS_{n_1,n_2,\ldots,n_r}$, where the vertex set is $V(CS_{n_1,n_2,\ldots,n_r}) = \{c_i : 1 \leq i \leq r\} \cup \bigcup_{i=1}^{r}\{u_i^j : 1 \leq j \leq n_i\}$, and the edge set is $E(CS_{n_1,n_2,\ldots n_r}) = \{c_i c_{i+1} : 1 \leq i \leq r-1\} \cup \{c_1 c_r\} \cup \bigcup_{i=1}^{r}\{c_i u_i^j : 1 \leq j \leq n_i\}$.

Theorem 13. *For closed caterpillar $CS_{n_1,n_2,\ldots,n_r}$ holds*

$$\text{lidis}(CS_{n_1,n_2,\ldots,n_r}) = \begin{cases} \infty, & \text{if } r = 3 \text{ and } \{n_1,n_2,n_3\} = \{n,0,0\}, \text{ where } n \geq 0, \\ 3, & \text{if } r = 3 \text{ and } (n_1,n_2,n_3) = (1,1,1), \\ 3, & \text{if } r = 3 + 6k, k \geq 1 \text{ and } \{n_1,n_2,\ldots,n_r\} = \{1,0,\ldots,0\}, \\ \leq 2, & \text{otherwise.} \end{cases}$$

The proof of the next result for the disjoint union of graphs, follows from the fact that there are no edges between the distinct components.

Theorem 14. *Let G_i, $i = 1,2,\ldots,m$ be arbitrary graphs. Then*

$$\text{lidis}\left(\bigcup_{i=1}^{m} G_i\right) = \max\{\text{lidis}(G_i) : i = 1,2,\ldots,m\}.$$

Immediately from the previous theorem, we obtain the following result.

Corollary 5. *Let n be a non-negative integer and let G be a graph. Then, $\text{lidis}(G \cup nK_1) = \text{lidis}(G)$.*

The *join* $G \oplus H$ of the disjoint graphs G and H is the graph $G \cup H$ together with all the edges joining vertices of $V(G)$ and vertices of $V(H)$. Let $\Delta(G)$ denote the maximal degree of the graph G.

Theorem 15. *For any graph G holds*

$$\text{lidis}(G \oplus K_1) = \begin{cases} \infty, & \text{if } \Delta(G) = |V(G)| - 1, \\ \text{lidis}(G), & \text{if } \Delta(G) < |V(G)| - 1. \end{cases}$$

Proof. Let w be the vertex of K_1. In a graph $G \oplus K_1$ the vertex w is adjacent to all vertices in G we immediately get that $\text{lidis}(G \oplus K_1) \geq \text{lidis}(G)$.

If $\Delta(G) = |V(G)| - 1$ then in $G \oplus K_1$ there are at least two vertices of degree $|V(G)| = |V(G \oplus K_1)| - 1$ and thus by Corollary 1 we have $\text{lidis}(G \oplus K_1) = \infty$.

Let $\Delta(G) < |V(G)| - 1$. If $\text{lidis}(G) = \infty$ then by Theorem 1 there exists at least two vertices, say u and v in G such that $N_G[u] = N_G[v]$. However, these vertices have the same closed neighborhood also in the graph $G \oplus K_1$ as

$$N_{G \oplus K_1}[u] = N_G[u] \cup \{w\} = N_G[v] \cup \{w\} = N_{G \oplus K_1}[v].$$

However, this implies that

$$\text{lidis}(G \oplus K_1) = \infty = \text{lidis}(G).$$

Now, consider that $\text{lidis}(G) < \infty$ and let f be a corresponding local inclusive distance vertex irregular graph of G. We define a labeling g of $G \oplus K_1$ in the following way

$$g(v) = \begin{cases} 1, & \text{if } v = w, \\ f(v), & \text{if } v \in V(G). \end{cases}$$

The induced vertex weights are

$$wt_g(v) = \begin{cases} \sum_{u \in V(G)} f(u) + 1, & \text{if } v = w, \\ wt_f(v) + 1, & \text{if } v \in V(G). \end{cases}$$

As $\Delta(G) < |V(G)| - 1$ we get that for any vertex $v \in V(G)$ is

$$wt_f(v) = \sum_{u \in N_G(v)} f(u) < \sum_{u \in V(G)} f(u).$$

Thus, all adjacent vertices have distinct weights. This means that g is a local inclusive distance vertex irregular labeling of $G \oplus K_1$. As vertex w is adjacent to every vertex in G we get $\text{lidis}(G \oplus K_1) = \text{lidis}(G)$ in this case. This concludes the proof. □

The graph in the previous theorem is not necessarily connected.

Theorem 16. *Let G_i, $i = 1, 2, \ldots, m$, $m \geq 2$ be arbitrary graphs. Then*

$$\text{lidis}\left(\left(\bigcup_{i=1}^{m} G_i\right) \oplus K_1\right) = \max\{\text{lidis}(G_i) : i = 1, 2, \ldots, m\}.$$

Proof. The proof follows from Theorems 14 and 15. □

A *wheel* W_n with n spokes is isomorphic to the graph $C_n \oplus K_1$. A *fan graph* F_n is isomorphic to the graph $P_n \oplus K_1$, while a *generalized fan graph* is isomorphic to the graph $kP_n \oplus K_1$. The following results are immediate corollaries of the previous theorems.

Corollary 6. *Let W_n be a wheel on $n + 1$ vertices $n \geq 3$. Then*

$$\text{lidis}(W_n) = \begin{cases} \infty, & \text{if } n = 3, \\ 2, & \text{if } n \text{ is even}, \\ 3, & \text{if } n \text{ is odd}, n \geq 5. \end{cases}$$

Corollary 7. *Let F_n be a fan on $n + 1$ vertices $n \geq 2$. Then*

$$\text{lidis}(F_n) = \begin{cases} \infty, & \text{if } n = 2, \\ 2, & \text{if } n \geq 3. \end{cases}$$

Corollary 8. *Let $kP_n \oplus K_1$ be a generalized fan graph, $k, n \geq 2$. Then*
$$\text{lidis}(kP_n \oplus K_1) = 2.$$

If $\text{lidis}(G) = \infty$ then by Theorem 1 there exist at least two vertices, say u and v in G such that they have the same closed neighborhood $N_G[u] = N_G[v]$. Thus, we immediately get

$$\begin{aligned} N_{G \oplus \overline{K_r}}[u] &= N_G[u] \cup \{w_i : i = 1, 2, \ldots, r\} \\ &= N_G[v] \cup \{w_i : i = 1, 2, \ldots, r\} = N_{G \oplus \overline{K_r}}[v], \end{aligned}$$

where w_i, $i = 1, 2, \ldots, r$, are the vertices of $\overline{K_r}$. Thus, $\text{lidis}(G \oplus \overline{K_r}) = \infty$ for every positive integer r. Now we will deal with the case when $\text{lidis}(G) < \infty$ and $r \geq 2$.

Theorem 17. *Let $r \geq 2$ be a positive integer and let G be not isomorphic to a totally disconnected graph. If $\text{lidis}(G) < \infty$ and $r \geq |V(G)| \cdot \text{lidis}(G)$ then $\text{lidis}(G \oplus \overline{K_r}) = \text{lidis}(G)$.*

Proof. Let us denote the vertices $\overline{K_r}$ by the symbols w_i, $i = 1, 2, \ldots, r$ and let $r \geq 2$. Thus, $V(G \oplus \overline{K_r}) = V(G) \cup \{w_i : i = 1, 2, \ldots, r\}$. In a graph $G \oplus \overline{K_r}$ all the vertices w_i, $i = 1, 2, \ldots, r$ are adjacent to all vertices in G thus we immediately get that $\text{lidis}(G \oplus \overline{K_r}) \geq \text{lidis}(G)$.

Let $\text{lidis}(G) < \infty$ and let f be a corresponding local inclusive distance vertex irregular labeling of G. We define a labeling g of $G \oplus \overline{K_r}$ in the following way

$$g(v) = \begin{cases} 1, & \text{if } v = w_i, i = 1, 2, \ldots, r, \\ f(v), & \text{if } v \in V(G). \end{cases}$$

Then, the vertex weights are

$$wt_g(v) = \begin{cases} \sum_{u \in V(G)} f(u) + 1, & \text{if } v = w_i, i = 1, 2, \ldots, r, \\ wt_f(v) + r, & \text{if } v \in V(G). \end{cases}$$

Evidently, under the labeling g, all adjacent vertices in $V(G)$ have distinct weights. We need also to prove that no vertex in $V(G)$ has the same weight as in $V(\overline{K_r})$. Consider that

$$r \geq |V(G)| \cdot \text{lidis}(G).$$

As G is not isomorphic to a totally disconnected graph then for the weight of any vertex v in $V(G)$ we have

$$wt_g(v) = wt_f(v) + r \geq 1 + |V(G)| \cdot \text{lidis}(G) > 1 + \sum_{u \in V(G)} f(u) = wt_g(w_i)$$

for every $i = 1, 2, \ldots, r$. Thus, g is a local inclusive distance vertex irregular graph of $G \oplus \overline{K_r}$ and hence $\text{lidis}(G \oplus \overline{K_r}) \leq \text{lidis}(G)$. □

Note that for small r the previous theorem is not necessarily true. Consider the graph G illustrated on Figure 2, evidently $\text{lidis}(G) = 1$. However, $\text{lidis}(G \oplus \overline{K_3}) = 2$.

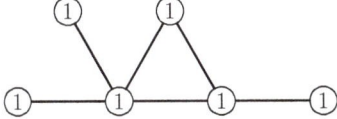

Figure 2. A local inclusive distance vertex irregular labeling of a graph G.

4. Conclusions

In this paper, we introduced the local inclusive distance vertex irregularity strength of graphs and gave some basic results and also some constructions of the feasible labelings for several families of graphs. We still have some open problems and conjecture as follows:

Problem 1. *Find* $\text{lidis}(K_{n_1,n_2,\ldots,n_m})$ *for general case, which is for the case* $n_1 \leq n_2 \leq \cdots \leq n_m$, *where* $m > 2$.

Problem 2. *Characterize graphs for which* $\text{lidis}(G \odot \overline{K_r}) = \text{lidis}(G)$.

Conjecture 1. *For arbitrary tree* T *with* $T \neq K_2$, $\text{lidis}(T) = 1$ *or* 2.

Author Contributions: Conceptualization, K.A.S., D.R.S., M.B. and A.S.-F.; methodology, K.A.S., D.R.S., M.B. and A.S.-F.; validation, K.A.S., D.R.S., M.B. and A.S.-F.; investigation, K.A.S., D.R.S., M.B. and A.S.-F.; resources, K.A.S., D.R.S., M.B. and A.S.-F.; writing—original draft preparation, K.A.S., and A.S.-F.; writing—review and editing, K.A.S., D.R.S., M.B. and A.S.-F.; supervision, K.A.S. and A.S.-F.; project administration, K.A.S., M.B. and A.S.-F.; funding acquisition, K.A.S., D.R.S., M.B. and A.S.-F. All authors have read and agreed to the published version of the manuscript.

Funding: This research has supported by PUTI KI-Universitas Indonesia 2020 Research Grant No. NKB-779/UN2.RST/HKP.05.00/2020. This work was also supported by the Slovak Research and Development Agency under the contract No. APVV-19-0153 and by VEGA 1/0233/18.

Institutional Review Board Statement: Not applicable.

Informed Consent Statement: Not applicable.

Data Availability Statement: Not applicable.

Conflicts of Interest: The authors declare no conflict of interest.

References

1. West, D.B. *Introduction to Graph Theory*, 2nd ed.; Prentice-Hall: Hoboken, NJ, USA, 2000.
2. Chartrand, G.; Jacobson, M.S.; Lehel, J.; Oellermann, O.R.; Ruiz, S.; Saba, F. Irregular networks. *Congr. Numer.* **1988**, *64*, 187–192.
3. Frieze, A.; Gould, R.J.; Karonski, M.; Pfender, F. On graph irregularity strength. *J. Graph Theory* **2002**, *41*, 120–137. [CrossRef]
4. Kalkowski, M.; Karonski, M.; Pfender, F. A new upper bound for the irregularity strength of graphs. *SIAM J. Discrete Math.* **2011**, *25*, 1319–1321. [CrossRef]
5. Majerski, P.; Przybyło, J. On the irregularity strength of dense graphs. *SIAM J. Discrete Math.* **2014**, *28*, 197–205. [CrossRef]
6. Miller, M.; Rodger, C.; Simanjuntak, R. Distance magic labelings of graphs. *Australas. J. Combin.* **2003**, *28*, 305–315.
7. Arumugam, S.; Fronček, D.; Kamatchi, N. Distance magic graphs—A survey. *J. Indones. Math. Soc.* **2011**, 11–26. [CrossRef]
8. Slamin, S. On distance irregular labelings of graphs. *Far East J. Math. Sci.* **2017**, *102*, 919–932.
9. Bong, N.H.; Lin, Y.; Slamin, S. On distance-irregular labelings of cycles and wheels. *Australas. J. Combin.* **2017**, *69*, 315–322.
10. Bong, N.H.; Lin, Y.; Slamin, S. On inclusive and non-inclusive vertex irregular d-distance vertex labelings. submitted.
11. Bača, M.; Semaničová-Feňovčíková, A.; Slamin, S.; Sugeng, K.A. On inclusive distance vertex irregular labelings. *Electron. J. Graph Theory Appl.* **2018**, *6*, 61–83. [CrossRef]
12. Utami, B.; Sugeng, K.A.; Utama, S. On inclusive d-distance irregularity strength on triangular ladder graph and path. *AKCE Int. J. Graphs Comb.* **2020**, *17*, 810–819. [CrossRef]
13. Utami, B.; Sugeng, K.A.; Utama, S. Inclusive vertex irregular 1-distance labelings on triangular ladder graphs. *AIP Conf. Proc.* **2021**, *2018*, 060006. [CrossRef]
14. Arumugam, S.; Premalatha, K.; Bača, M.; Semaničová-Feňovčíková, A. Local antimagic vertex coloring of a graph. *Graphs Combin.* **2017**, *33*, 275–285. [CrossRef]
15. Czerwiński, S.; Grytczuk, J.; Żelazny, W. Lucky labelings of graphs. *Inform. Process. Lett.* **2009**, *109*, 1078–1081. [CrossRef]
16. Gallian, J.A. A dynamic survey of graph labeling. *Electron. J. Combin.* **2019**, #DS6. Available online: https://www.combinatorics.org/ojs/index.php/eljc/article/view/DS6/pdf (accessed on 13 July 2021)

Article

On the Quasi-Total Roman Domination Number of Graphs

Abel Cabrera Martínez [1], Juan C. Hernández-Gómez [2] and José M. Sigarreta [2,*]

[1] Departament d'Enginyeria Informàtica i Matemàtiques, Universitat Rovira i Virgili, Av. Països Catalans 26, 43007 Tarragona, Spain; abel.cabrera@urv.cat
[2] Facultad de Matemáticas, Universidad Autónoma de Guerrero, Carlos E. Adame No. 54, Col. Garita, Acapulco 39650, Mexico; jcarloshg@gmail.com
* Correspondence: josemariasigarretaalmira@hotmail.com

Abstract: Domination theory is a well-established topic in graph theory, as well as one of the most active research areas. Interest in this area is partly explained by its diversity of applications to real-world problems, such as facility location problems, computer and social networks, monitoring communication, coding theory, and algorithm design, among others. In the last two decades, the functions defined on graphs have attracted the attention of several researchers. The Roman-dominating functions and their variants are one of the main attractions. This paper is a contribution to the Roman domination theory in graphs. In particular, we provide some interesting properties and relationships between one of its variants: the quasi-total Roman domination in graphs.

Keywords: quasi-total Roman domination; total Roman domination; Roman domination

1. Introduction

Domination in graphs was first defined as a graph-theoretical concept in 1958. This area has attracted the attention of many researchers due to its diversity of applications to real-world problems, such as problems with the location of facilities, computing and social networks, communication monitoring, coding theory, and algorithm design, among others. In that regard, this topic has experienced rapid growth, resulting in over 5000 papers being published. We refer to [1,2] for theoretical results and practical applications.

Given a graph G, a *dominating set* is a subset $D \subseteq V(G)$ of vertices, such that every vertex not in D is adjacent to at least one vertex in D. The minimum cardinality among all dominating sets of G is called the *domination number* of G. The number of works, results and open problems that exist on this parameter and its variants provide a very wide range of work directions to consider, which come from their very theoretical aspects to a significant number of practical applications, passing through a large number of relationships and connections between some invariants of graph theory itself.

In the last two decades, the interest in research concerning dominating functions in graphs has increased. One of the reasons for this is that dominating functions generalize the concept of dominating sets. In particular, the Roman dominating functions (defined in [3], due to historical reasons arising from the ancient Roman Empire and described in [4,5]), and their variants, are one of the main attractions. At present, more than 300 papers have been published on this topic.

In 2019, Cabrera García et al. [6] defined and began the study of an interesting variant of Roman-dominating functions: the quasi-total Roman-dominating functions. This paper deals precisely with this style of domination, and our goal is to continue with the study of this novel parameter in graphs.

Definitions, Notation and Organization of the Paper

We begin this subsection by stating the main basic terminology which will be used in the whole work. Let $G = (V(G), E(G))$ be a simple graph with no isolated vertex. Given a vertex $v \in V(G)$, $N(v) = \{x \in V(G) : xv \in E(G)\}$ and $N[v] = N(v) \cup \{v\}$. A vertex $v \in V(G)$ is called a *leaf vertex* if $|N(v)| = 1$. Given a set $D \subseteq V(G)$, $N(D) = \cup_{v \in D} N(v)$, $N[D] = N(D) \cup D$ and $\partial(D) = N(D) \setminus D$. Moreover, given a set $D \subseteq V(G)$ and a vertex $v \in D$, $epn(v, D) = \{u \in V(G) \setminus D : N(u) \cap D = \{v\}\}$. Also, and as is commonly defined, $G - D$ denotes the graph obtained from G such that $V(G - D) = V(G) \setminus D$ and $E(G - D) = E(G) \setminus \{uv \in E(G) : u \in D$ or $v \in D\}$. Moreover, the subgraph of G induced by $D \subseteq V(G)$ will be denoted by $G[D]$.

We say that G is *F-free* if it contains no copy of F as an induced subgraph. A set $D \subseteq V(G)$ is a *2-packing* if $N[x] \cap N[y] \neq \emptyset$ for every pair $x, y \in D$. The *2-packing number* of G, denoted by $\rho(G)$, is defined as $\max\{|D| : D$ is a 2-packing of $G\}$. A 2-packing of cardinality $\rho(G)$ is called a $\rho(G)$-set. We will assume an analogous correspondence when referring to the optimal sets or functions derived from other parameters used in the article.

Let $f : V(G) \to \{0, 1, 2\}$ be a function on G. Observe that f generates three sets V_0, V_1 and V_2, where $V_i = \{v \in V(G) : f(v) = i\}$ for $i \in \{0, 1, 2\}$. In this sense, we will write $f(V_0, V_1, V_2)$ to refer to the function f. Given a set $D \subseteq V(G)$, $f(D) = \sum_{v \in D} f(v)$. The *weight* of f is defined as $\omega(f) = f(V(G)) = |V_1| + 2|V_2|$. We shall also use the following notations: $V_{1,2} = \{v \in V_1 : N(v) \cap V_2 \neq \emptyset\}$, $V_{1,0} = \{v \in V_1 : N(v) \subseteq V_0\}$ and $V_{1,1} = V_1 \setminus (V_{1,2} \cup V_{1,0})$. A function $f(V_0, V_1, V_2)$ on G is a *dominating function* if $N(v) \cap (V_1 \cup V_2) \neq \emptyset$ for every vertex $v \in V_0$. Moreover, f is a *total dominating function* (TDF) if $N(v) \cap (V_1 \cup V_2) \neq \emptyset$ for every vertex $v \in V(G)$. Next, we highlight some particular cases of known domination parameters, which we define here in terms of (total) dominating functions.

- A set $D \subseteq V(G)$ is a *(total) dominating set* of G if there exists a (total) dominating function $f(V_0, V_1, V_2)$ such that $f(x) > 0$ if, and only if, $x \in D$. The *(total) domination number* of G, denoted by $(\gamma_t(G))$ $\gamma(G)$, is the minimum cardinality among all (total) dominating sets of G. For more information on domination and total domination see the books [1,2,7], the survey [8] and the recent works [9–11].
- A function $f(V_0, V_1, V_2)$ is a *Roman-dominating function* if $N(v) \cap V_2 \neq \emptyset$ for every $v \in V_0$. The *Roman domination number* of G, denoted by $\gamma_R(G)$, is the minimum weight among all Roman-dominating functions on G. For more information on Roman domination and its varieties, see the articles [3,12].
- A TDF $f(V_0, V_1, V_2)$ is a *total Roman-dominating function* (TRDF) on a graph G without isolated vertices if $N(v) \cap V_2 \neq \emptyset$ for every vertex $v \in V_0$. The *total Roman domination number*, denoted by $\gamma_{tR}(G)$, is the minimum weight among all TRDFs on G. For recent results on the total Roman domination in graphs we cite [13–20].
- A *quasi-total Roman-dominating function* (QTRDF) on a graph G is a dominating function $f(V_0, V_1, V_2)$ such that $N(x) \cap V_2 \neq \emptyset$ for every $x \in V_0$; and $N(y) \cap (V_1 \cup V_2) \neq \emptyset$ for every $y \in V_2$. The minimum weight among all QTRDFs on G is called the *quasi-total Roman domination number*, and is denoted by $\gamma_{qtR}(G)$. This parameter was introduced by Cabrera Martínez et al. [6].

As consequence of the above definitions and the well-known inequalities $\rho(G) \leq \gamma(G)$ (see [1]), $\gamma_t(G) \leq \gamma_R(G)$ (see [21]) and $\gamma_{tR}(G) \leq \gamma_R(G) + \gamma(G)$ (see [14]), we establish an inequality chain involving the previous parameters.

Theorem 1. *If G is a graph with no isolated vertex, then*

$$\rho(G) \leq \gamma(G) \leq \gamma_t(G) \leq \gamma_R(G) \leq \gamma_{qtR}(G) \leq \gamma_{tR}(G) \leq \gamma_R(G) + \gamma(G).$$

For instance, for the graphs G_1 and G_2 given in Figure 1 we deduce the next inequality chains. In that regard, the labels of (gray and black) coloured vertices describe the positive weights of a $\gamma_{qtR}(G_i)$-function, for $i \in \{1, 2\}$.

- $\rho(G_1) = 1 < 3 = \gamma(G_1) < 4 = \gamma_t(G_1) < 5 = \gamma_R(G_1) < 6 = \gamma_{qtR}(G_1) < 7 = \gamma_{tR}(G_1)$.
- $\rho(G_2) = 1 < 3 = \gamma(G_2) < 4 = \gamma_t(G_2) < 6 = \gamma_R(G_2) < 7 = \gamma_{qtR}(G_2) = \gamma_{tR}(G_2)$.

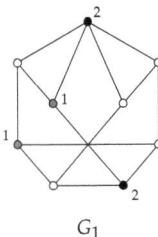 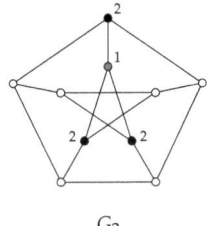

Figure 1. The labels of (gray and black) coloured vertices describe the positive weights of a $\gamma_{qtR}(G_i)$-function, for $i \in \{1, 2\}$.

As mentioned before, the goal of this work is continue the study of the quasi-total Roman domination number of graphs. In that regard, the paper is organized as follows. First, we obtain new, tight bounds for this parameter. Such bounds can also be seen as relationships between this novel parameter and several other classical domination parameters such as the (total) domination and (total) Roman domination numbers. Finally, and as a consequence of this previous study, we derive new results on the total Roman domination number of a graph.

2. Bounds and Relationships with Other Parameters

Let G be a disconnected graph and let $G_1, \ldots, G_r$ ($r \geq 2$) be the components of G. Observe that any QTRDF $f(V_0, V_1, V_2)$ on G satisfies that f restricted to $V(G_j)$ is a QTRDF on G_j, for every $j \in \{1, \ldots, r\}$. Therefore, the following result is obtained for the case of disconnected graphs.

Remark 1 ([6])**.** *Let $G_1, \ldots, G_r$ ($r \geq 2$) be the components of a disconnected graph G. Then*

$$\gamma_{qtR}(G) = \sum_{i=1}^{r} \gamma_{qtR}(G_i).$$

As a consequence of the above remark, throughout this paper, we only consider nontrivial connected graphs. Next, we give two useful lemmas, which provide some tools to deduce some of the results.

Lemma 1. *Let G be a nontrivial connected graph. If $f(V_0, V_1, V_2)$ is a $\gamma_{qtR}(G)$-function, then the following statements hold.*

(i) *$f'(V_0' = V_0, V_1' = V_1 \setminus V_{1,0}, V_2' = V_2)$ is a $\gamma_{tR}(G - V_{1,0})$-function.*
(ii) *$epn(v, V_2) \cap V_0 \neq \emptyset$, for every $v \in V_2$.*
(iii) *If $\gamma_{qtR}(G) = \gamma_R(G)$, then $V_{1,2} = \emptyset$.*

Proof. Let $f(V_0, V_1, V_2)$ be a $\gamma_{qtR}(G)$-function. First, we proceed to prove (i). Notice that $G - V_{1,0}$ has no isolated vertex. Hence, the function $f'(V_0', V_1', V_2')$, defined by $V_0' = V_0$, $V_1' = V_1 \setminus V_{1,0}$ and $V_2' = V_2$, is a TRDF on $G - V_{1,0}$. Hence, $\gamma_{tR}(G - V_{1,0}) \leq \omega(f')$. Now, if $\gamma_{tR}(G - V_{1,0}) < \omega(f')$, then from any $\gamma_{tR}(G - V_{1,0})$-function and the set $V_{1,0}$, we can construct a QTRDF on G of weight less than $\omega(f) = \gamma_{qtR}(G)$, which is a contradiction. Therefore, the function f' is a $\gamma_{tR}(G - V_{1,0})$-function, as required.

Now, we proceed to prove (ii). Let $v \in V_2$. Obviously, $N(v) \cap V_0 \neq \emptyset$. If $epn(v, V_2) \cap V_0 = \emptyset$, then the function $f'(V_0', V_1', V_2')$, defined by $V_1' = V_1 \cup \{v\}$, $V_2' = V_2 \setminus \{v\}$ and $V_0' = V_0$, is a QTRDF on G of weight $\omega(f') < \omega(f) = \gamma_{qtR}(G)$, which is a contradiction. Therefore, $epn(v, V_2) \cap V_0 \neq \emptyset$, which completes the proof of (ii).

Finally, we proceed to prove (iii). Assume that $\gamma_{qtR}(G) = \gamma_R(G)$. First, suppose that $V_{1,2} \neq \emptyset$. It is easy to see that the function $f'(V_0', V_1', V_2')$, defined by $V_1' = V_1 \setminus V_{1,2}$, $V_2' = V_2$ and $V_0' = V_0 \cup V_{1,2}$, is a Roman-dominating function on G. Hence, $\gamma_R(G) \leq \omega(f') <$

$\omega(f) = \gamma_{qtR}(G)$, which is a contradiction. Therefore, $V_{1,2} = \emptyset$, which completes the proof of (iii). □

Lemma 2. *Let G be a nontrivial connected graph. If $f(V_0, V_1, V_2)$ is a $\gamma_{qtR}(G)$-function such that $|V_1|$ is minimum, then one of the following conditions holds.*

(i) $V_{1,0} = \emptyset$.
(ii) $V_{1,0}$ *is a 2-packing of G.*

Proof. Let $f(V_0, V_1, V_2)$ be a $\gamma_{qtR}(G)$-function, such that $|V_1|$ is minimal. Assume that $V_{1,0} \neq \emptyset$. It is clear by definition that $V_{1,0}$ is an independent set. Now, suppose that $V_{1,0}$ is not a 2-packing of G. Therefore, two vertices $u, v \in V_{1,0}$ exist at distance two. Let $w \in N(u) \cap N(v)$. Notice that $w \in V_0$ and $N(w) \cap V_2 \neq \emptyset$. With these conditions in mind, observe that the function $f'(V'_0, V'_1, V'_2)$, defined by $V'_1 = V_1 \setminus \{u, v\}$, $V'_2 = V_2 \cup \{w\}$ and $V'_0 = V(G) \setminus (V'_1 \cup V'_2)$, is a QTRDF on G of weight $\omega(f') = \omega(f)$ and $|V'_1| < |V_1|$, which is a contradiction. Therefore, $V_{1,0}$ is a 2-packing of G, which completes the proof. □

We continue with one of the main results of this paper.

Theorem 2. *If G is a nontrivial connected graph, then at least one of the following statements holds.*

(i) $\gamma_{qtR}(G) = \gamma_{tR}(G)$.
(ii) $\gamma_{qtR}(G) = \min\{\gamma_{tR}(G - S) + |S| : S \text{ is a 2-packing of } G\}$.

Proof. Let $f(V_0, V_1, V_2)$ be a $\gamma_{qtR}(G)$-function such that $|V_1|$ is minimum. If $V_{1,0} = \emptyset$, then by Lemma 1-(i) we deduce that f is also a $\gamma_{tR}(G)$-function, which implies that $\gamma_{qtR}(G) = \gamma_{tR}(G)$. Hence, from now on, we assume that $V_{1,0} \neq \emptyset$. By Lemma 2, it follows that $V_{1,0}$ is a 2-packing of G. Moreover, by Lemma 1-(i) we have the function $f'(V'_0 = V_0, V'_1 = V_1 \setminus V_{1,0}, V'_2 = V_2)$ is a $\gamma_{tR}(G - V_{1,0})$-function. Therefore, $\gamma_{qtR}(G) = \gamma_{tR}(G - V_{1,0}) + |V_{1,0}| \geq \min\{\gamma_{tR}(G - S) + |S| : S \text{ is a 2-packing of } G\}$. We only need to prove that $\gamma_{qtR}(G) \leq \min\{\gamma_{tR}(G - S) + |S| : S \text{ is a 2-packing of } G\}$. In such a sense, let S be a 2-packing of G for which $\gamma_{tR}(G - S) + |S|$ is minimum, and let $g'(W'_0, W'_1, W'_2)$ be a $\gamma_{tR}(G - S)$-function. Observe that the function $g(W_0, W_1, W_2)$, defined by $W_0 = W'_0$, $W_1 = W'_1 \cup S$ and $W_2 = W'_2$, is a QTRDF on G. Therefore, $\gamma_{qtR}(G) \leq \omega(g) = \min\{\gamma_{tR}(G - S) + |S| : S \text{ is a 2-packing of } G\}$, which completes the proof. □

The next proposition is a direct consequence of Theorem 2.

Proposition 1. *If G is a nontrivial connected graph, then*

$$\gamma_{qtR}(G) \geq \gamma_{tR}(G) - \rho(G).$$

Proof. If $\gamma_{qtR}(G) = \gamma_{tR}(G)$, then the inequality holds. Assume that $\gamma_{qtR}(G) < \gamma_{tR}(G)$. By Theorem 2 there exists a 2-packing S of G such that $\gamma_{qtR}(G) = \gamma_{tR}(G - S) + |S|$. Let $f'(V'_0, V'_1, V'_2)$ be a $\gamma_{tR}(G - S)$-function and let $S' \subseteq N(S)$ be a set of cardinality $|S|$ such that $N(x) \cap S' \neq \emptyset$ for every vertex $x \in S$. Observe that the function $f(V_0, V_1, V_2)$, defined by $V_2 = V'_2$, $V_1 = V'_1 \cup S \cup (S' \setminus V'_2)$ and $V_0 = V(G) \setminus (V_1 \cup V_2)$, is a TRDF on G. Therefore, $\gamma_{tR}(G) \leq \omega(f) \leq \omega(f') + |S| + |S'| = \gamma_{tR}(G - S) + 2|S| = \gamma_{qtR}(G) + |S| \leq \gamma_{qtR}(G) + \rho(G)$, which completes the proof. □

The bound above is tight. For instance, it is achieved for the graph G given in the Figure 2. Notice that this figure describes the positive weights of a $\gamma_{qtR}(G)$-function. In addition, it is easy to see that $\rho(G) = 2$ and $\gamma_{tR}(G) = 8$. Hence, $\gamma_{qtR}(G) = 6 = \gamma_{tR}(G) - \rho(G)$, as required.

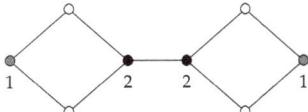

Figure 2. The labels of (gray and black) coloured vertices describe the positive weights of a $\gamma_{qtR}(G)$-function.

It is well-known that $\gamma_{tR}(G) \geq 2\gamma(G) \geq \gamma_R(G)$ for any graph G with no isolated vertex (see [3,15]). From this inequality chain, we deduce the following result.

Theorem 3. *For any nontrivial connected graph G,*
$$2\gamma(G) - \rho(G) \leq \gamma_{qtR}(G) \leq 3\gamma(G).$$

Proof. By combining the bound $\gamma_{tR}(G) \geq 2\gamma(G)$ and the bound given in Proposition 1, we deduce that $\gamma_{qtR}(G) \geq 2\gamma(G) - \rho(G)$.

Now, from the bound $\gamma_R(G) \leq 2\gamma(G)$ and the inequality $\gamma_{qtR}(G) \leq \gamma_R(G) + \gamma(G)$ given in Theorem 1 we obtain $\gamma_{qtR}(G) \leq \gamma_R(G) + \gamma(G) \leq 3\gamma(G)$, as desired. □

The lower bounds given in the two previous results are tight. We will show later that, as a consequence of Lemma 3, the graphs $G_{a,0} \in \mathcal{G}$ satisfy the equality established in Proposition 1, while the graph $G_{2,0}$ satisfies the equality given in Theorem 3.

With respect to the equality in the bound $\gamma_{qtR}(G) \leq 3\gamma(G)$ above, we can see that this bound is tight. For instance, it is achieved for the graph G given in the Figure 3. Notice that this figure describes the positive weights of a $\gamma_{qtR}(G)$-function, and as a consequence, we deduce that $\gamma_{qtR}(G) = 6 = 3\gamma(G)$, as required.

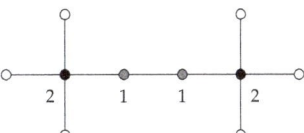

Figure 3. The labels of (gray and black) coloured vertices describe the positive weights of a $\gamma_{qtR}(G)$-function.

In addition, we can deduce the following connection. To this end, we need to say that a graph G is called a *Roman graph* if $\gamma_R(G) = 2\gamma(G)$.

Proposition 2. *If G is a graph such that $\gamma_{qtR}(G) = 3\gamma(G)$, then G is a Roman graph.*

Proof. From the proof of Theorem 3, we have that $3\gamma(G) = \gamma_{qtR}(G) \leq \gamma_R(G) + \gamma(G) \leq 3\gamma(G)$. Thus, we have equalities in the inequality chain above. In particular, $\gamma_R(G) = 2\gamma(G)$, which completes the proof. □

Notice that the opposed to the proposition above is not necessarily true. For instance, the graph G_2 given in Figure 1 is a Roman graph, but it does not satisfy the equality $\gamma_{qtR}(G_2) = 3\gamma(G_2)$.

The following result gives a lower bound for the quasi-total Roman domination number and characterizes the class of connected graphs for which $\gamma_{qtR}(G) \in \{\gamma(G) + 1, \gamma(G) + 2\}$.

Theorem 4. *For any nontrivial connected graph G of order n,*
$$\gamma_{qtR}(G) \geq \gamma(G) + 1.$$

Furthermore,

(i) $\gamma_{qtR}(G) = \gamma(G) + 1$ *if and only if* $G \cong P_2$.
(ii) $\gamma_{qtR}(G) = \gamma(G) + 2$ *if and only if one of the following conditions holds.*

   (a) $G \not\cong P_2$ *has a vertex of degree* $n - \gamma(G)$.
   (b) G *has two adjacent vertices* u, v *such that* $|\partial(\{u, v\})| = n - \gamma(G)$.

Proof. If $G \cong P_2$, then it is clear that $\gamma_{qtR}(G) = \gamma(G) + 1$. From now on, assume that $G \not\cong P_2$. Let $f(V_0, V_1, V_2)$ be a $\gamma_{qtR}(G)$-function, such that $|V_1|$ is minimum. Note that $(V_1 \setminus V_{1,2}) \cup V_2$ is a dominating set of G, and $|V_2| \geq 1$. Hence, $\gamma_{qtR}(G) = 2|V_2| + |V_1| \geq (|V_2| + |V_1 \setminus V_{1,2}|) + |V_2| \geq \gamma(G) + 1$, and the lower bound follows.

Now, suppose that $\gamma_{qtR}(G) = \gamma(G) + 1$. So, we have equalities in the inequality chain above. In particular, $V_{1,2} = \emptyset$ and $|V_2| = 1$, which is a contradiction. Therefore, if $G \not\cong P_2$, then $\gamma_{qtR}(G) \geq \gamma(G) + 2$, and, as a consequence, (i) follows.

We next proceed to prove (ii). First, suppose that $\gamma_{qtR}(G) = \gamma(G) + 2$. Notice that,

$$\gamma(G) + 2 = \omega(f) \geq (|V_2| + |V_1 \setminus V_{1,2}|) + |V_2| \geq \gamma(G) + |V_2|.$$

This implies that $|V_2| \in \{1, 2\}$, and, in such a case, we consider the following two cases.

Case 1. $|V_2| = 1$. In this case, we have that $|V_1| = \gamma(G)$. Let $V_2 = \{v\}$. Now, as $|N(v) \cap V_1| = 1$ and $V_0 \subseteq N(v)$, we deduce that $|N(v)| = |V_0| + 1 = (n - |V_1| - |V_2|) + 1 = n - \gamma(G)$, which implies that condition (a) follows.

Case 2. $|V_2| = 2$. Let $V_2 = \{u, v\}$. In this case we have that $|V_1| = \gamma(G) - 2$, and we have equalities in the inequality chain above. As a consequence, $V_{1,2} = \emptyset$, which implies that u and v are adjacent vertices. Hence, $\partial(\{u, v\}) = V_0$ and, therefore, $|\partial(\{u, v\})| = |V_0| = n - |V_1| - |V_2| = n - \gamma(G)$. Therefore, condition (b) follows.

On the other hand, suppose that one of the conditions (a) and (b) holds. In such a sense, we consider the next two cases. Recall that $\gamma_{qtR}(G) \geq \gamma(G) + 2$ since $G \not\cong P_2$.

Case 1. Suppose that (a) holds. Let $v \in V(G)$ such that $|N(v)| = n - \gamma(G)$ and $w \in N(v)$. Notice that the function $f'(V'_0, V'_1, V'_2)$, defined by $V'_2 = \{v\}$, $V'_0 = N(v) \setminus \{w\}$ and $V'_1 = V(G) \setminus (V'_0 \cup V'_2)$, is a QTRDF on G. Hence, $\gamma_{qtR}(G) \leq \omega(f') = 2|V'_2| + |V'_1| = 2 + \gamma(G)$, which implies that $\gamma_{qtR}(G) = \gamma(G) + 2$, as required.

Case 2. Suppose that (b) holds. Let u, v be two adjacent vertices such that $|\partial(\{u, v\})| = n - \gamma(G)$. Observe that the function $f''(V''_0, V''_1, V''_2)$, defined by $V''_2 = \{u, v\}$, $V''_0 = \partial(\{u, v\})$ and $V''_1 = V(G) \setminus (V''_0 \cup V''_2)$, is a QTRDF on G. Hence, $\gamma_{qtR}(G) \leq \omega(f'') = 2|V''_2| + |V''_1| = 4 + (\gamma(G) - 2) = \gamma(G) + 2$, which implies that $\gamma_{qtR}(G) = \gamma(G) + 2$, as required. Therefore, the proof is complete. □

Cabrera Martínez et al. [6] in 2019, established that $\gamma_{qtR}(G) \leq n - \rho(G)(\delta(G) - 2)$ for any nontrivial graph G of order n and minimum degree $\delta(G)$. The following bounds for the total Roman domination number and the domination number, respectively, are direct consequences of the previous inequality, Proposition 1 and Theorem 3.

Theorem 5. *The following statements hold for any nontrivial connected graph* G *of order* n *and* $\delta(G) \geq 4$.

(i) $\gamma_{tR}(G) \leq n - \rho(G)(\delta(G) - 3)$.
(ii) $\gamma(G) \leq \frac{n - \rho(G)(\delta(G) - 3)}{2}$.

From Proposition 1 and Theorem 1, we obtain the following useful inequality chain.

$$\gamma_{tR}(G) - \rho(G) \leq \gamma_{qtR}(G) \leq \gamma_{tR}(G). \tag{1}$$

An interesting question that arises from the inequality chain above is the following. Can the differences $\gamma_{qtR}(G) - (\gamma_{tR}(G) - \rho(G))$ and $\gamma_{tR}(G) - \gamma_{qtR}(G)$ be as large as possible? Next, we provide an affirmative answer to the previous question. For this purpose, we need to introduce the following family of graphs. Given two integers $a, b \geq 0$ $(a + b \geq 2)$, a graph $G_{a,b} \in \mathcal{G}$ is defined as follows.

- We begin with a nontrivial connected graph G of order $|V(G)| = a + b$ with vertex set $V(G) = \{u_1, \ldots, u_a, v_1, \ldots, v_b\}$.
- Attach a path $P_4 = x_1 x_2 x_3 x_4$ to every $u_i \in V(G), i \in \{1, \ldots, a\}$, by adding an edge between u_i and every vertex in $\{x_1, x_2, x_4\}$.
- Attach a double star $S_{1,2}$ to every $v_j \in V(G), j \in \{1, \ldots, b\}$, by adding an edge between v_j and every leaf vertex of $S_{1,2}$.

The Figure 4 shows the graph $G_{2,3}$ by taking $G \cong P_5$. We next give exact formulas for the total Roman domination number, the quasi-total Roman domination number and the packing number of the graphs of the family $\mathcal{G}$. These results are almost straightforward to deduce and, according to this fact, the proofs are left to the reader.

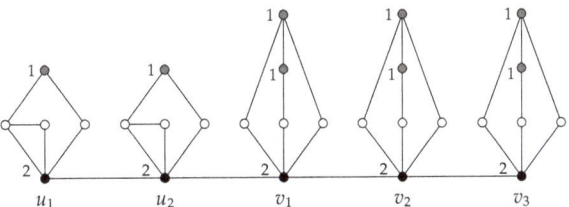

Figure 4. The graph $G_{2,3}$ by taking G as the path graph P_5. The labels of (gray and black) coloured vertices describe the positive weights of a $\gamma_{qtR}(G_{2,3})$-function.

Lemma 3. Let $a, b \geq 0$ be two integers, such that $a + b \geq 2$. If G is a connected graph such that $|V(G)| = a + b$, then the following equalities hold.

(i) $\gamma_{tR}(G_{a,b}) = 4a + 4b$.
(ii) $\gamma_{qtR}(G_{a,b}) = 3a + 4b$.
(iii) $\rho(G_{a,b}) = a + b$.

According to the lemma above, for any integers $a, b \geq 0$ $(a + b \geq 2)$, we obtain that any graph $G_{a,b} \in \mathcal{G}$ satisfies

$$\gamma_{qtR}(G_{a,b}) - (\gamma_{tR}(G_{a,b}) - \rho(G_{a,b})) = b \quad \text{and} \quad \gamma_{tR}(G_{a,b}) - \gamma_{qtR}(G_{a,b}) = a,$$

which provides the answer to our previous question. In addition, and as a consequence of Lemma 3, we deduce that the lower and upper bounds given in Inequality chain (1) are tight. For instance, any graph $G_{a,0} \in \mathcal{G}$ satisfies that $\gamma_{qtR}(G_{a,0}) = \gamma_{tR}(G_{a,0}) - \rho(G_{a,0})$, while any graph $G_{0,b} \in \mathcal{G}$ satisfies that $\gamma_{qtR}(G_{0,b}) = \gamma_{tR}(G_{0,b})$.

It is well known that $\rho(G) = 1$ for every graph G with a diameter of, at most, two. In this sense, and as direct consequence of the Inequality chain (1), we have that $\gamma_{qtR}(G) \in \{\gamma_{tR}(G) - 1, \gamma_{tR}(G)\}$ for every graph G with diameter of, at most, two. We next show some subclasses which satisfy the equality $\gamma_{qtR}(G) = \gamma_{tR}(G)$. For this, we need to cite the following result.

Theorem 6 ([6]). *The following statements hold for any nontrivial graph G.*

(i) $\gamma_{qtR}(G) = 2$ if and only if $G \cong P_2$.
(ii) $\gamma_{qtR}(G) = 3$ if and only if $G \not\cong P_2$ and $\gamma(G) = 1$.
(iii) $\gamma_{qtR}(G) = 4$ if and only if $\gamma_t(G) = \gamma(G) = 2$.

The join of two graphs G_1 and G_2, denoted by $G_1 + G_2$, is the graph obtained from G_1 and G_2 with vertex set $V(G_1 + G_2) = V(G_1) \cup V(G_2)$ and edge set $E(G_1 + G_2) = E(G_1) \cup E(G_2) \cup \{uv : u \in V(G_1), v \in V(G_2)\}$. Observe that $diam(G_1 + G_2) \leq 2$ by definition.

The following result, which is a consequence of Theorem 6, shows that $\gamma_{tR}(G_1 + G_2) = \gamma_{qtR}(G_1 + G_2)$.

Theorem 7. *For any nontrivial graphs G_1 and G_2,*

$$\gamma_{qtR}(G_1 + G_2) = \gamma_{tR}(G_1 + G_2) = \begin{cases} 3, & \text{if } \min\{\gamma(G_1), \gamma(G_2)\} = 1; \\ 4, & \text{otherwise.} \end{cases}$$

We continue analysing other subclasses of graphs with a diameter of two. The following results consider the planar graphs with a diameter of two.

Theorem 8 ([22]). *If G is a planar graph with $diam(G) = 2$, then the following statements hold.*
(i) $\gamma(G) \leq 2$ or $G = G_9$, where G_9 is the graph given in Figure 5.
(ii) $\gamma_t(G) \leq 3$.

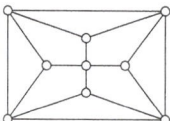

Figure 5. The planar graph G_9 with $diam(G_9) = 2$ and $\gamma_t(G_9) = \gamma(G_9) = 3$.

Theorem 9. *For any planar graph G with $diam(G) = 2$,*

$$\gamma_{qtR}(G) = \gamma_{tR}(G) = \begin{cases} 3, & \text{if } \gamma(G) = 1; \\ 4, & \text{if } \gamma(G) = \gamma_t(G) = 2; \\ 5, & \text{if } \gamma_t(G) = \gamma(G) + 1 = 3; \\ 6, & \text{if } G = G_9. \end{cases}$$

Proof. If $G = G_9$, then it is easy to check that $\gamma_{qtR}(G) = \gamma_{tR}(G) = 6$. From now on, let $G \neq G_9$ be a planar graph with $diam(G) = 2$. It is straightforward that $\gamma_{qtR}(G) = \gamma_{tR}(G) = 3$ if and only if $\gamma(G) = 1$. Hence, assume that $\gamma(G) \geq 2$. By Theorem 8, it follows that $\gamma(G) = 2$ and $\gamma_t(G) \in \{2, 3\}$. Next, we analyse these two cases.

Case 1. $\gamma_t(G) = 2$. By Theorems 6 and 1 and the well-known bound $\gamma_{tR}(G) \leq 2\gamma_t(G)$ (see [15]) we obtain that $4 = \gamma_{qtR}(G) \leq \gamma_{tR}(G) \leq 2\gamma_t(G) = 4$. Thus, $\gamma_{qtR}(G) = \gamma_{tR}(G) = 4$.

Case 2. $\gamma_t(G) = 3$. As a consequence of the Theorem 6 we have that $\gamma_{qtR}(G) \geq 5$. Let $\{u, v\}$ be a $\gamma(G)$-set. Since $\gamma_t(G) = 3$ and $diam(G) = 2$, it follows that u and v are at distance two. Let $w \in N(u) \cap N(v)$. Notice that the function f, defined by $f(u) = f(v) = 2$, $f(w) = 1$ and $f(x) = 0$ for every $x \in V(G) \setminus \{u, v, w\}$, is a TRDF on G, which implies that $\gamma_{tR}(G) \leq w(f) = 5$. Hence, by the fact that $\gamma_{qtR}(G) \leq \gamma_{tR}(G)$ we deduce that $\gamma_{qtR}(G) = \gamma_{tR}(G) = 5$.

Therefore, the proof is complete. □

However, for the case of non-planar graphs with a diameter of two, there are graphs that satisfy $\gamma_{qtR}(G) = \gamma_{tR}(G)$ or $\gamma_{qtR}(G) = \gamma_{tR}(G) - 1$. For instance, for the graphs G_1 and G_2 given in Figure 1 we have that $\gamma_{qtR}(G_1) = 6 = \gamma_{tR}(G_1) - 1$ and $\gamma_{qtR}(G_2) = 7 = \gamma_{tR}(G_2)$. In connection with this fact, we pose the following open problem.

Problem 1. *Characterize the families of non-planar graphs G with diameter two for which $\gamma_{qtR}(G) = \gamma_{tR}(G)$ or $\gamma_{qtR}(G) = \gamma_{tR}(G) - 1$.*

Notice that, as consequence of the Inequality chain (1), any new result for the total Roman domination number gives us a new result for the quasi-total Roman domination number and vice versa. In such a sense, we continue with two new bounds for the total Roman domination number. Before this, we need to introduce the following definition.

A set S of vertices of a graph G is a *vertex cover* if every edge of G is incident with at least one vertex in S. The *vertex cover number* of G, denoted by $\beta(G)$, is the minimum cardinality among all vertex covers of G.

Theorem 10. *For any $K_{1,3}$-free graph G with $\delta(G) \geq 3$,*

$$\gamma_{tR}(G) \leq \beta(G) + \gamma(G).$$

Proof. Let D be a $\gamma(G)$-set and S a $\beta(G)$-set. Let $f(V_0, V_1, V_2)$ be a function defined by $V_0 = V(G) \setminus (D \cup S)$, $V_1 = (D \cup S) \setminus (D \cap S)$ and $V_2 = D \cap S$. Now, we proceed to prove that f is a TRDF on G. We first note that S is a total dominating set because G is $K_{1,3}$-free. Hence, $V_1 \cup V_2 = D \cup S$ is a total dominating set of G. Let $v \in V_0 = V(G) \setminus (D \cup S)$. So, $N(v) \subseteq S$ and $N(v) \cap D \neq \emptyset$. Hence $N(v) \cap D \cap S \neq \emptyset$, i.e., $N(v) \cap V_2 \neq \emptyset$. Therefore, f is a TRDF on G, as desired. Thus, $\gamma_{tR}(G) \leq \omega(f) \leq |(D \cup S) \setminus (D \cap S)| + 2|D \cap S| = \beta(G) + \gamma(G)$, which completes the proof. □

Lemma 4 ([15]). *If G is a graph with no isolated vertex, then there exists a $\gamma_{tR}(G)$-function $f(V_0, V_1, V_2)$ such that either V_2 is a dominating set of G, or the set S of vertices not dominated by V_2 satisfies $G[S] = kK_2$ for some $k \geq 1$, where $S \subseteq V_1$ and $\partial(S) \subseteq V_0$.*

Theorem 11. *If G is a $\{K_{1,3}, K_{1,3} + e\}$-free graph such that $\delta(G) \geq 3$, then there exists a $\gamma_{tR}(G)$-function $f(V_0, V_1, V_2)$ such that V_2 is a dominating set of G, and, as a consequence,*

$$\gamma_{tR}(G) \geq \gamma_t(G) + \gamma(G).$$

Proof. Suppose that there is no $\gamma_{tR}(G)$-function $f(V_0', V_1', V_2')$ such that V_2' is a dominating set of G. By Lemma 4, there exists a $\gamma_{tR}(G)$-function $f(V_0, V_1, V_2)$ such that $V_{1,1}$ satisfies that $G[V_{1,1}] = kK_2$ for some $k \geq 1$ and $\partial(V_{1,1}) \subseteq V_0$. We can assume that $|V_1|$ is minimum among all $\gamma_{tR}(G)$-functions because it is a requirement for the existence of the function f (see the proof of Lemma 4). Let $u, v \in V_{1,1}$ be two adjacent vertices. Hence, $\partial(\{u,v\}) \subseteq V_0$. Since $\delta(G) \geq 3$, there are two vertices $w_1, w_2 \in N(v) \cap V_0$, and as G is a $\{K_{1,3}, K_{1,3} + e\}$-free graph, we deduce that at least one of these vertices is also adjacent to u. Hence, and without loss of generality, assume that $\{u,v\} \subseteq N(w_1)$. Observe that the function $g(W_0, W_1, W_2)$, defined by $W_2 = V_2 \cup \{w_1\}$, $W_1 = V_1 \setminus \{u,v\}$ and $W_0 = V(G) \setminus (W_1 \cup W_2)$, is a TRDF on G of weight $\omega(g) = \omega(f)$ and $|W_1| < |V_1|$, which is a contradiction. Therefore, there exists a $\gamma_{tR}(G)$-function $f(V_0, V_1, V_2)$ such that V_2 is a dominating set of G. Since $V_1 \cup V_2$ is a total dominating set of G, we deduce that $\gamma_t(G) + \gamma(G) \leq |V_1 \cup V_2| + |V_2| = 2|V_2| + |V_1| = \gamma_{tR}(G)$, which completes the proof. □

Observe that, if G is a $\{K_{1,3}, K_{1,3} + e\}$-free graph of minimum degree at least three with $\beta(G) = \gamma_t(G)$, then the bounds given in the two previous theorems are achieved. Moreover, let G be a $(n-2)$-regular graph obtained from the complete graph K_n (n even) by deleting the edges of a perfect matching. Notice that G is $\{K_{1,3}, K_{1,3} + e\}$-free and satisfies that $\gamma_{tR}(G) = 4 = \gamma_t(G) + \gamma(G)$.

Theorem 12. *If G is a connected $\{K_{1,3}, K_{1,3} + e\}$-free graph such that $\delta(G) \geq 3$, then the following statements hold.*

(i) $\gamma_t(G) + \gamma(G) - \rho(G) \leq \gamma_{qtR}(G) \leq \beta(G) + \gamma(G)$.
(ii) *If $\gamma_{tR}(G) = \gamma_R(G)$, then $\gamma_{qtR}(G) = 2\gamma_t(G)$.*

Proof. Statement (i) is a direct consequence of combining Inequality chain (1) and Theorems 10 and 11. Finally, we proceed to prove (ii). By Theorem 11 there exists a $\gamma_{tR}(G)$-function $f(V_0, V_1, V_2)$ such that V_2 is a dominating set of G. Hence, $V_{1,1} = \emptyset$. Moreover, as $\gamma_{tR}(G) = \gamma_R(G)$, we deduce that f is also a $\gamma_{qtR}(G)$-function and Lemma 1 (iii)-(a) leads to $V_{1,2} = \emptyset$. Therefore $V_1 = \emptyset$, which implies that V_2 is a total dominating set of G. Hence, $2\gamma_t(G) \leq 2|V_2| = \gamma_{qtR} = \gamma_{tR} \leq 2\gamma_t(G)$. Therefore, $\gamma_{qtR} = 2\gamma_t(G)$, as required. □

3. Conclusions and Open Problems

This paper is a contribution to the graph domination theory. We have studied the quasi-total Roman domination in graphs. For instance, we have shown the close relationship that exists between this novel parameter and other invariants, such as (total) domination number, (total) Roman domination number and 2-packing number.

We conclude by proposing some open problems.

- Settle Problem 1.
- Characterize the graphs that satisfy the following equalities.
    - $\gamma_{qtR}(G) = \gamma_{tR}(G)$.
    - $\gamma_{qtR}(G) = \gamma_{tR}(G) - \rho(G)$.
    - $\gamma_{qtR}(G) = 3\gamma(G)$.
- We have shown that if G is a $\{K_{1,3}, K_{1,3} + e\}$-free graph with minimum degree $\delta(G) \geq 3$, then $\gamma_{qtR}(G) \geq \gamma_t(G) + \gamma(G) - \rho(G)$. We conjecture that the previous bound holds for any graph with no isolated vertex.

Author Contributions: Investigation, A.C.M., J.C.H.-G. and J.M.S. All authors contributed equally to this work. All authors have read and agreed to the published version of the manuscript.

Funding: This research was funded by Agencia Estatal de Investigación grant number PID2019-106433GB-I00/AEI/10.13039/501100011033, Spain.

Institutional Review Board Statement: Not applicable.

Informed Consent Statement: Not applicable.

Conflicts of Interest: The authors declare no conflict of interest.

References

1. Haynes, T.W.; Hedetniemi, S.T.; Slater, P.J. *Fundamentals of Domination in Graphs*; Marcel Dekker, Inc.: New York, NY, USA, 1998.
2. Haynes, T.W.; Hedetniemi, S.T.; Slater, P.J. *Fundamentals of Domination in Graphs: Advanced Topics*; Chapman & Hall, CRC Press: Boca Raton, FL, USA, 1998.
3. Cockayne, E.J.; Dreyer, P.A.; Hedetniemi, S.M.; Hedetniemi, S.T. Roman domination in graphs. *Discret. Math.* **2004**, *278* 11–22. [CrossRef]
4. Revelle, C.S.; Rosing, K.E. Defendens imperium romanum: A classical problem in military strategy. *Am. Math. Mon.* **2000**, *107*, 585–594. [CrossRef]
5. Stewart, I. Defend the Roman Empire. *Sci. Am.* **1999**, *28*, 136–139. [CrossRef]
6. Cabrera, G.S.; Cabrera, M.A.; Yero, I.G. Quasi-total Roman domination in graphs. *Results Math.* **2019**, *74*, 173. [CrossRef]
7. Henning, M.A.; Yeo, A. Total Domination in Graphs. In *Monographs in Mathematics*; Springer: Berlin, Germany, 2013; ISBN: 978-1461465249.
8. Henning, M.A. A survey of selected recent results on total domination in graphs. *Discret. Math.* **2009**, *309*, 32–63. [CrossRef]
9. Henning, M.A.; Yeo, A. A new upper bound on the total domination number in graphs with minimum degree six. *Discret. Appl. Math.* **2021**, *302*, 1–7. [CrossRef]
10. Sigarreta, J.M. Total domination on some graph operators. *Mathematics* **2021**, *9*, 241. [CrossRef]
11. Cabrera, M.A.; Rodríguez-Velázquez, J.A. Total domination in rooted product graphs. *Symmetry* **2020**, *12*, 1929. [CrossRef]
12. Chellali, M.; Jafari Rad, N.; Sheikholeslami, S.M.; Volkmann, L. Varieties of Roman domination II. *AKCE Int. J. Graphs Comb.* **2020**, *17*, 966–984. [CrossRef]
13. Liu, C.-H.; Chang, G.J. Roman domination on strongly chordal graphs. *J. Comb. Optim.* **2013**, *26*, 608–619. [CrossRef]
14. Cabrera, M.A.; Cabrera, G.S.; Carrión, G.A. Further results on the total Roman domination in graphs. *Mathematics* **2020**, *8*, 349. [CrossRef]
15. Abdollahzadeh, A.H.; Henning, M.A.; Samodivkin, V.; Yero, I.G. Total Roman domination in graphs. *Appl. Anal. Discrete Math.* **2016**, *10*, 501–517. [CrossRef]
16. Cabrera, M.A.; Cabrera, G.S.; Carrión, G.A.; Hernéz, M.F.A. Total Roman domination number of rooted product graphs. *Mathematics* **2020**, *8*, 1850. [CrossRef]
17. Amjadi, J.; Sheikholeslami, S.M.; Soroudi, M. Nordhaus-Gaddum bounds for total Roman domination. *J. Comb. Optim.* **2018**, *35*, 126–133. [CrossRef]

18. Cabrera, M.A.; Kuziak, D.; Peterin, I.; Yero, I.G. Dominating the direct product of two graphs through total Roman strategies. *Mathematics* **2020**, *8*, 1438.
19. Cabrera, M.A.; Rodríguez-Velázquez, J.A. Closed formulas for the total Roman domination number of lexicographic product graphs. *Ars Math. Contemp.* **2021**, in press. [CrossRef]
20. Campanelli, N.; Kuziak, D. Total Roman domination in the lexicographic product of graphs. *Discret. Appl. Math.* **2019**, *263*, 88–95. [CrossRef]
21. Chellali, M.; Haynes, T.W.; Hedetniemi, S.T. Roman and total domination. *Quaest. Math.* **2015**, *38*, 749–757. [CrossRef]
22. Goddard, W.; Henning, M.A. Domination in planar graphs with small diameter. *J. Graph Theory* **2002**, *40*, 1–25. [CrossRef]

Article

On the Double Roman Domination in Generalized Petersen Graphs $P(5k, k)$

Darja Rupnik Poklukar [1] and Janez Žerovnik [1,2,*]

[1] Faculty of Mechanical Engineering, University of Ljubljana, Aškerčeva 6, 1000 Ljubljana, Slovenia; darja.rupnik-poklukar@fs.uni-lj.si
[2] Institute of Mathematics, Physics and Mechanics, Jadranska 19, 1000 Ljubljana, Slovenia
* Correspondence: janez.zerovnik@fs.uni-lj.si

Abstract: A double Roman dominating function on a graph $G = (V, E)$ is a function $f : V \to \{0, 1, 2, 3\}$ satisfying the condition that every vertex u for which $f(u) = 0$ is adjacent to at least one vertex assigned 3 or at least two vertices assigned 2, and every vertex u with $f(u) = 1$ is adjacent to at least one vertex assigned 2 or 3. The weight of f equals $w(f) = \sum_{v \in V} f(v)$. The double Roman domination number $\gamma_{dR}(G)$ of a graph G equals the minimum weight of a double Roman dominating function of G. We obtain closed expressions for the double Roman domination number of generalized Petersen graphs $P(5k, k)$. It is proven that $\gamma_{dR}(P(5k, k)) = 8k$ for $k \equiv 2, 3 \mod 5$ and $8k \leq \gamma_{dR}(P(5k, k)) \leq 8k + 2$ for $k \equiv 0, 1, 4 \mod 5$. We also improve the upper bounds for generalized Petersen graphs $P(20k, k)$.

Keywords: double Roman domination; generalized Petersen graph; discharging method; graph cover; double Roman graph

---

Citation: Rupnik Poklukar, D.; Žerovnik, J. On the Double Roman Domination in Generalized Petersen Graphs $P(5k, k)$. *Mathematics* **2022**, *10*, 119. https://doi.org/10.3390/math10010119

Academic Editor: Mikhail Goubko

Received: 25 November 2021
Accepted: 24 December 2021
Published: 1 January 2022

Publisher's Note: MDPI stays neutral with regard to jurisdictional claims in published maps and institutional affiliations.

Copyright: © 2022 by the authors. Licensee MDPI, Basel, Switzerland. This article is an open access article distributed under the terms and conditions of the Creative Commons Attribution (CC BY) license (https://creativecommons.org/licenses/by/4.0/).

1. Introduction

Double Roman domination of graphs was first studied in [1], motivated by a number of applications of Roman domination in present time and in history [2]. The initial studies of Roman domination [3,4] have been motivated by a historical application. In the 4th century, Emperor Constantine was faced with a difficult problem of how to defend the Roman Empire with limited resources. His decision was to allocate two types of armies to the provinces in such a way that all the provinces in the empire will be safe. Some military units were well trained and capable of moving rapidly from one city to another in order to respond to any attack. Other legions consisted of a local militia and they were permanently positioned in a given province. The Emperor decreed that no legion could ever leave a province to defend another if in this case they left the province undefended. Thus, at some provinces two units were stationed, a local militia units were stationed at others, and some provinces had no army. While the problem is still of interest in military operations research [5], it also has applications in cases where a time-critical service is to be provided with some backup. For example, a fire station should never send all emergency vehicles to answer a call.

Similar reasoning applies in any emergency service. Hence positioning the fire stations, first aid stations, etc. at optimal positions improves the public services without increasing the cost. A natural generalization, in particular in the case of emergency services, is the k-Roman domination [6], where in the district not one, but k emergency teams are expected to be quickly available in case of multiple emergency calls. Special case $k = 2$, the double Roman domination, is considered in this work. It is well-known that the decision version of the double Roman domination problem (MIN-DOUBLE-RDF) is NP-complete, even when restricted to planar graphs, chordal graphs, bipartite graphs, undirected path graphs, chordal bipartite graphs and to circle graphs [7–9]. It is therefore of interest to study the complexity of the problem for other families of graphs. For example, linear time algorithms exist for interval graphs and block graphs [8], for trees [10], for proper interval graphs [11] and for unicyclic graphs [9]. Another avenue of research that is motivated by high complexity of the problem is to obtain closed expressions for the double Roman domination number of some families of graphs. In particular, generalized Petersen graphs and certain subfamilies of generalized Petersen graphs have been studied extensively in recent years. The results listed in subsection on related previous work include closed expressions for the double Roman domination number of some, and tight bounds for other subfamilies [12–15]. For more results on double Roman domination we refer to recent papers [16–19] and the references there.

The rest of the paper is organized as follows. In Section 2, we recall some basic definitions and known results that will be used in the following sections. In the last part of Section 2 our main results, Theorems 6 and 7, are presented. In Section 3, we present upper and lower bounds for double Roman domination number in generalized Petersen graphs $P(5k,k)$. Finally, in Section 4, we give an improved upper bounds for double Roman domination number of generalized Petersen graphs $P(20k,k)$, using the notion of covering graphs.

2. Preliminaries

2.1. Graphs and Double Roman Domination

Let $G = (V, E)$ be a graph without loops and multiple edges. As usual, we denote with $V = V(G)$ the vertex set of G and with $E = E(G)$ its edge set.

A set $D \subseteq V(G)$ is a dominating set if every vertex in $V(G) \setminus D$ has at least one neighbor in D. The domination number $\gamma(G)$ is the cardinality of a minimum dominating set of G. A double Roman dominating function (DRDF) on a graph $G = (V, E)$ is a function $f : V \to \{0, 1, 2, 3\}$ with the following properties:

(1) every vertex u with $f(u) = 0$ is adjacent to at least one vertex assigned 3 or at least two vertices assigned 2, and
(2) every vertex u with $f(u) = 1$ is adjacent to at least one vertex assigned 2 or 3 under f.

Define $f(U) = \sum_{u \in U} f(u)$ as the weight of f on an arbitrary subset $U \subseteq V(G)$. Then, the weight of f equals $w(f) = f(V(G)) = \sum_{v \in V(G)} f(v)$. The double Roman domination number $\gamma_{dR}(G)$ of a graph G is the minimum weight of a double Roman dominating function of G. A DRD function f is called a γ_{dR}-function of G if $w(f) = \gamma_{dR}(G)$.

For any double Roman dominating function f, defined on G we define a partition of the vertex set $V = V_0 \cup V_1 \cup V_2 \cup V_3$, where $V_i = V_i^f = \{u \mid f(u) = i\}$.

The study of the double Roman domination in graphs was initiated by Beeler et al. [1]. It was proved that $2\gamma(G) \leq \gamma_{dR}(G) \leq 3\gamma(G)$. Furthermore, Beeler at al. defined a graph G to be double Roman if $\gamma_{dR}(G) = 3\gamma(G)$, where $\gamma(G)$ is the domination number of G. For a later reference we recall the following result, also obtained by Beeler et al.

Proposition 1 ([1]). *In a double Roman dominating function f of weight $\gamma_{dR}(G)$, no vertex needs to be assigned the value 1.*

Domination in graphs with its many varieties has been extensively studied in the past [20–23]. Roman domination and double Roman domination is a rather new variety of interest [1,2,7,24–27].

2.2. Generalized Petersen Graphs

The generalized Petersen graph $P(n,k)$ is a graph with vertex set $U \cup V$ and edge set $E_1 \cup E_2 \cup E_3$, where $U = \{u_0, u_1, \cdots, u_{n-1}\}$, $V = \{v_0, v_1, \cdots, v_{n-1}\}$, $E_1 = \{u_i u_{i+1} \mid i = 0, 1, \ldots, n-1\}$, $E_2 = \{u_i v_i \mid i = 0, 1, \ldots, n-1\}$, $E_3 = \{v_i v_{i+k} \mid i = 0, 1, \ldots, n-1\}$, and subscripts are reduced modulo n, see Figure 1. Thus, we identify integers i and j iff $i \equiv j \mod n$. (As usual, $m \equiv r \mod n$ means that $m = kn + r$, or equivalently, $m - r = kn$ for some integer $k \in \mathbb{Z}$.)

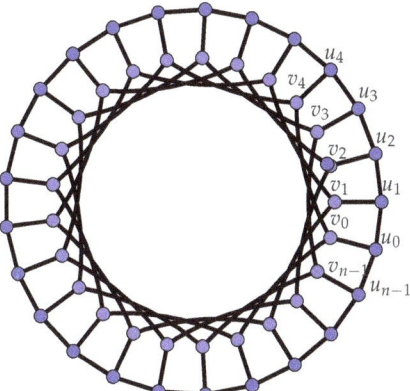

Figure 1. A generalized Petersen graph $P(n, k)$.

It is well known that the graphs $P(n, k)$ are 3-regular unless $k = \frac{n}{2}$ and that $P(n, k)$ are highly symmetric [28,29]. As $P(n, k)$ and $P(n, n - k)$ are isomorphic, it is natural to restrict attention to $P(n, k)$ with $n \geq 3$ and k, $1 \leq k < \frac{n}{2}$.

Petersen graphs are among the most interesting examples when considering nontrivial graph invariants. The domination and its variations (such as vertex domination, exact domination, rainbow domination, double Roman domination and other) of generalized Petersen graphs have been extensively studied in recent years, see for example [14,30–36].

2.3. Related Previous Work

The domination number for the generalized Petersen graphs $P(ck, k)$ for integer constants $c \geq 3$ was studied by Zhao et al. [37]. They obtained upper bound on $\gamma(P(ck, k))$ for general c.

Theorem 1 ([37]). *For any $k \geq 1$ and $c \geq 3$*

$$\gamma(P(ck,k)) \leq \begin{cases} \frac{c}{3} \left\lceil \frac{5k}{3} \right\rceil, & c \equiv 0 \bmod 3, \\ \left\lceil \frac{c}{3} \right\rceil \left\lceil \frac{5k}{3} \right\rceil - \left\lceil \frac{2k}{3} \right\rceil, & c \equiv 1 \bmod 3, \\ \left\lceil \frac{c}{3} \right\rceil \left\lceil \frac{5k}{3} \right\rceil - \left\lceil \frac{2k}{3} \right\rceil + \left\lceil \frac{k}{3} \right\rceil, & c \equiv 2 \bmod 3. \end{cases}$$

Shao et al. [14] determine the exact value of $\gamma_{dR}(P(n,1))$, and Jiang et al. [13] determine $\gamma_{dR}(P(n,2))$.

Theorem 2 ([13,14]). *Let $n \geq 3$. Then we have*

$$\gamma_{dR}(P(n,1)) = \begin{cases} \frac{3n}{2}, & n \equiv 0 \bmod 4, \\ \frac{3n+3}{2}, & n \equiv 1, 3 \bmod 4, \\ \frac{3n+4}{2}, & n \equiv 2 \bmod 4 \end{cases}$$

and for $n \geq 5$

$$\gamma_{dR}(P(n,2)) = \begin{cases} \left\lceil \frac{8n}{5} \right\rceil, & n \equiv 0 \bmod 5, \\ \left\lceil \frac{8n}{5} \right\rceil + 1, & n \equiv 1,2,3,4 \bmod 5. \end{cases}$$

Shao et al. in [14] obtained also a general lower bound on double Roman domination numbers for arbitrary graphs of a maximum degree greater or equal one.

Theorem 3 ([14]). *If G is a graph of maximum degree $\triangle \geq 1$, then*

$$\gamma_{dR}(G) \geq \left\lceil \frac{3V(G)}{\triangle+1} \right\rceil.$$

Clearly, as the generalized Petersen graph $P(n,k)$ is 3-regular and has $2n$ vertices.

Corollary 1 ([14]). *In Petersen graphs $P(n,k)$, $\gamma_{dR}(P(n,k)) \geq \lceil \frac{3n}{2} \rceil$.*

Gao et al. [12] determined the exact value of $\gamma_{dR}(P(n,k))$ for $n \equiv 0 \mod 4$ and $k \equiv 1 \mod 2$, and presented an improved upper bound for $\gamma_{dR}(P(n,k))$ in other cases. The results are summarized in the next theorem.

Theorem 4 ([12]). *For $k \geq 3$, $\gamma_{dR}(P(n,k)) = \frac{3n}{2}$, $k \equiv 1 \mod 2$, $n \equiv 0 \mod 4$.*

$$\left\lceil \frac{3n}{2} \right\rceil \leq \gamma_{dR}(P(n,k)) \leq \begin{cases} \frac{3n}{2} + \frac{5k+5}{4}, & k \equiv 1 \mod 4, n \not\equiv 0 \mod 4, \\ \frac{3n}{2} + \frac{5k+7}{4}, & k \equiv 3 \mod 4, n \not\equiv 0 \mod 4, \\ \frac{3n}{2} \frac{(3k+2)}{(3k+1)}, & k \equiv 0 \mod 4, n \equiv 0 \mod (3k+1), \\ \lceil \frac{3n}{2} \frac{(3k+2)}{(3k+1)} \rceil + \frac{5k+4}{4}, & k \equiv 0 \mod 4, n \not\equiv 0 \mod (3k+1), \\ \frac{3n}{2} \frac{(3k)}{(3k-1)}, & k \equiv 2 \mod 4, n \equiv 0 \mod (3k-1), \\ \lceil \frac{3n}{2} \frac{(3k)}{(3k-1)} \rceil + \frac{5k+6}{4}, & k \equiv 2 \mod 4, n \not\equiv 0 \mod (3k-1). \end{cases}$$

Double Roman domination of families $P(ck,k)$ has been studied recently for small k, including $c = 3, 4$, and 5. Shao et al. [15] considered the double Roman domination number in generalized Petersen graphs $P(3k,k)$ and constructed solutions providing the upper bounds, which gives exact values for $\gamma_{dR}(P(3k,k))$.

Theorem 5 ([15]).

$$\gamma_{dR}(P(3k,k)) = \begin{cases} 5k+1, & k \in \{1,2,4\} \\ 5k, & \text{otherwise} \end{cases}$$

For small cases in the families $P(4k,k)$ and $P(5k,k)$, the known facts are summarized in the next proposition.

Proposition 2. *$\gamma_{dR}(P(4,1)) = 6$, $\gamma_{dR}(P(8,2)) = 14$ [13], $\gamma_{dR}(P(12,3)) = 18$ and $\gamma_{dR}(P(5,1)) = 9$ [14], $\gamma_{dR}(P(10,2)) = 16$ [13], $23 \leq \gamma_{dR}(P(15,3)) \leq 26$ [12].*

Wang et al. in [38] showed that $\gamma(P(4k,k)) = \begin{cases} 2k; & k \equiv 1 \mod 2, \\ 2k+1; & k \equiv 0 \mod 2 \end{cases}$, and $\gamma(P(5k,k)) = 3k$ for all $k \geq 1$. Furthermore, recall the lower bound given in Corollary 1 and recall that Theorem 4 implies $\gamma_{dR}(P(4k,k)) = 6k$ for $k \equiv 1 \mod 2$. Thus, we can write the known facts regarding $\gamma_{dR}(P(4k,k))$ and $\gamma_{dR}(P(5k,k))$ in the next two propositions.

Proposition 3. *Let $k \geq 1$. If $k \equiv 1 \mod 2$, then $\gamma_{dR}(P(4k,k)) = 6k$, and if $k \equiv 0 \mod 2$ then $6k \leq \gamma_{dR}(P(4k,k)) \leq 6k+3$.*

Proposition 4. *Let $k \geq 3$. Then $7k + \lceil \frac{k}{2} \rceil \leq \gamma_{dR}(P(5k,k)) \leq 9k$.*

2.4. Our Results

The main result of our paper are either exact values or narrow bounds for the double Roman domination numbers of all Petersen graphs $P(5k, k)$. More precisely, we will show that the following theorem holds.

Theorem 6. *Let $k \geq 2$.*

$$8k \leq \gamma_{dR}(P(5k,k)) \leq \begin{cases} 8k, & k \equiv 2,3 \bmod 5 \\ 8k+2, & otherwise \end{cases}$$

As mentioned earlier, a graph G is double Roman if $\gamma_{dR}(G) = 3\gamma(G)$. Using the known equality $\gamma(P(5k,k)) = 3k$ for all $k \geq 1$ [38], we can conclude that the only double Roman graph in the set of generalized Petersen graphs $P(5k,k)$ is $P(5,1)$.

Corollary 2. *There is no double Roman graphs in the set of generalized Petersen graphs $P(5k,k)$ for $k \geq 2$. The graph $P(5,1)$ is a double Roman graph.*

We also show that certain generalized Petersen graphs are covering graphs of other generalized Petersen graphs (Proposition 7). This provides a method for establishing new upper bounds, see Proposition 8. In particular, we elaborate the case $P(20k, k)$ to obtain exact values in some, and tight bounds in other cases.

Theorem 7. $\gamma_{dR}(P(20,1)) \leq 40$, $\gamma_{dR}(P(40,2)) \leq 64$, $\gamma_{dR}(P(60,3)) \leq 96$. *Furthermore, let $k > 3$. Then, for odd k, we have*

$$\gamma_{dR}(P(20k,k)) = 30k \tag{1}$$

and for k even,

$$30k \leq \gamma_{dR}(P(20k,k)) \leq 30k + 15. \tag{2}$$

By Proposition 1, we can only consider the DRDF of a graph G with no vertex assigned the value 1.

3. Constructions and Proofs

In this section, the constructions of double Roman dominating functions providing upper bounds for the double Roman dominating numbers are given. We start by introducing some convenient notation for representing the DRDFs and providing some basic constructions.

In order to present the double Roman dominating functions of generalized Petersen graphs as concise as possible we use two different notations. For smaller graphs, we use the notation in brackets, showing weights on outer and inner cycles in two lines:

$$\begin{pmatrix} f(u_0) & f(u_1) & \ldots & f(u_{n-1}) \\ f(v_0) & f(v_1) & \ldots & f(v_{n-1}) \end{pmatrix}.$$

For example, a DRD function showing $\gamma_{dR}(P(4,1)) = 6$ is the following:

$$\begin{pmatrix} 0 & 0 & 3 & 0 \\ 3 & 0 & 0 & 0 \end{pmatrix}.$$

For bigger graphs we use the notation that provides only the values on the outer cycle, see Table 1. (In this case, the assignment on the inner cycles is completed such that the weight is minimal, see Lemma 1 for details.)

The columns correspond to the sets $U_i = \{u_i, u_{i+k}, u_{i+2k}, \ldots, u_{i+(c-1)k}\}$, and we assume that the inner cycles, sets $V_i = \{v_i, v_{i+k}, v_{i+2k}, \ldots, v_{i+(c-1)k}\}$, are completed such that the whole assignment presents a DRD function. As we can see from Table 1 below, the first two and the last two columns provide the same information on DRD function, namely the values at $U_0 = U_k$, and $U_1 = U_{k+1}$. We will use this property observing certain patterns appearing in case of optimal assignment —it will hold exactly when columns 0 and k will match, taking into account the shift of rows as indicated in Table 1.

Table 1. A DRD function of U_i for $P(4k, k)$.

$f(u_0)$	$f(u_1)$	...	$f(u_i)$	...	$f(u_{k-1})$	$f(u_k)$	$f(u_{k+1})$	...
$f(u_k)$	$f(u_{k+1})$	...	$f(u_{k+i})$	...	$f(u_{2k-1})$	$f(u_{2k})$	$f(u_{2k+1})$	...
$f(u_{2k})$	$f(u_{2k+1})$	...	$f(u_{2k+i})$	...	$f(u_{3k-1})$	$f(u_{3k})$	$f(u_{3k+1})$	...
$f(u_{3k})$	$f(u_{3k+1})$	...	$f(u_{3k+i})$	...	$f(u_{4k-1})$	$f(u_{4k}) = f(u_0)$	$f(u_1)$	...
0	1	...	i	...	$k-1$	k	$k+1$	...

To better understand the notation in tables, consider the pattern in Table 2 that provides DRDF(double Roman dominating function) for $P(12, 3)$ and $P(28, 7)$.

Table 2. An optimal DRDF(double Roman dominating function) of U_i for $P(4k, k)$. The first column provides a DRD function for $P(4, 1)$, the first 3 columns provide a DRD function for $P(12, 3)$, and the first 7 columns provide a DRD function for $P(28, 7)$.

0	0	3	0	0	0	3	0	...
0	0	0	3	0	0	0	3	...
3	0	0	0	3	0	0	0	...
0	3	0	0	0	3	0	0	...
0	1	2	3	4	5	6	7	...

Considering closely the graph $P(12, 3)$ and using the fact that the columns 0 and 4 correspond to the same set of vertices, $U_0 = U_4$, and that the column 4 equals column 0 shifted one row downwards (see Table 1), we can see that the pattern is well defined on the outer cycle of $P(12, 3)$. Obviously, the vertices on the inner cycles could be assigned with three more weights of 3, so we have a DRDF of $P(12, 3)$ of weight $9 + 9 = 18$. Similarly, regarding $P(28, 7)$, we have $U_0 = U_7$, and the same reasoning applies. Recalling Theorem 4, the constructions are best possible (compare the bounds in Theorem 3.)

3.1. Basic Constructions for $P(5k, k)$.

Recall that $\gamma_{dR}(P(5, 1)) = 9$ [14]. A simple DRD function showing $\gamma_{dR}(P(5, 1)) \leq 9$ is the following:

$$\begin{pmatrix} 3 & 0 & 0 & 0 & 3 \\ 0 & 0 & 3 & 0 & 0 \end{pmatrix}.$$

For larger graphs among $P(5k, k)$, we are going to use the notation introduced in Table 3.

Table 3. A DRD function of U_i for $P(5k, k)$.

$f(u_0)$	$f(u_1)$	...	$f(u_i)$	...	$f(u_{k-1})$	$f(u_k)$	$f(u_{k+1})$	...
$f(u_k)$	$f(u_{k+1})$	...	$f(u_{k+i})$	...	$f(u_{2k-1})$	$f(u_{2k})$	$f(u_{2k+1})$	...
$f(u_{2k})$	$f(u_{2k+1})$	...	$f(u_{2k+i})$	...	$f(u_{3k-1})$	$f(u_{3k})$	$f(u_{3k+1})$	...
$f(u_{3k})$	$f(u_{3k+1})$	...	$f(u_{3k+i})$	...	$f(u_{4k-1})$	$f(u_{4k})$	$f(u_{4k+1})$	...
$f(u_{4k})$	$f(u_{4k+1})$	...	$f(u_{4k+i})$	...	$f(u_{5k-1})$	$f(u_{5k}) = f(u_0)$	$f(u_1)$	...
0	1	...	i	...	$k-1$	k	$k+1$	...

The next two examples provide constructions of DRDF that show $\gamma_{dR}(P(10,2)) \leq 16$ and $\gamma_{dR}(P(35,7)) \leq 56$.

We proceed with two comments on Table 4.

- First, note that the columns in Table 4 have the following properties: each column has two consecutive vertices that are assigned two legions: say u_i and u_{i+k} for some i. Then, in the column i, vertices u_{i+2k} and u_{i+4k} have one neighbor and the vertex u_{i+3k} has two neighbors in the outer cycle that are assigned 2. Clearly, the missing legions can be provided by assigning weight 2 to vertices v_{i+2k} and v_{i+4k} on the inner cycle. In this assignment, each of the vertices u_{i+2k}, u_{i+3k} and u_{i+4k} is adjacent to two vertices of weight 2. Hence we have weight 8 for each column.
- Second, observe that columns 2 and 7 coincide. Recalling the convention given in Table 3, note that for $k = 2$, column 2 is column 0 shifted one row upwards (cyclicaly). Similarly, for $k = 7$, and columns 0 and 7.

Table 4. A DRDF of U_i that implies $\gamma_{dR}(P(10,2)) \leq 16$ and $\gamma_{dR}(P(35,7)) \leq 56$.

2	0	2	0	0	2	0	2	...
2	0	0	2	0	2	0	0	...
0	2	0	2	0	0	2	0	...
0	2	0	0	2	0	2	0	...
0	0	2	0	2	0	0	2	...
0	1	2	3	4	5	6	7	8	...

These constructions are optimal, which will follow from the lower bound that will be proved below. Recall that $\gamma_{dR}(P(10,2)) = 16$ [13], therefore the RDF for $\gamma_{dR}(P(10,2))$ given in Table 4 is best possible.

A symmetrical construction, given in Table 5 shows $\gamma_{dR}(P(15,3)) \leq 24$ and $\gamma_{dR}(P(40,8)) \leq 64$.

Table 5. A DRDF of U_i that implies $\gamma_{dR}(P(15,3)) \leq 24$ and $\gamma_{dR}(P(40,8)) \leq 64$.

2	0	0	2	0	2	0	0	2	...
2	0	2	0	0	2	0	2	0	...
0	0	2	0	2	0	0	2	0	...
0	2	0	0	2	0	2	0	0	...
0	2	0	2	0	0	2	0	2	...
0	1	2	3	4	5	6	7	8	...

3.2. Double Roman Domination in $P(5k,k)$—Upper Bounds

Observe that the patterns used in Tables 4 and 5 have period five (columns). This implies the next proposition.

Proposition 5. *Let $k \equiv 2 \bmod 5$ or $k \equiv 3 \bmod 5$. Then $\gamma_{dR}(P(5k,k)) \leq 8k$.*

Proof. Recall from previous considerations (Table 4) that $\gamma_{dR}(P(10,2)) \leq 16$ and $\gamma_{dR}(P(35,7)) \leq 56$. Observe that if we repeat the columns 2-6 in Table 4, we obtain a DRDF showing $\gamma_{dR}(P(60,12)) \leq 96$. By induction, it follows that $\gamma_{dR}(P(5(5i+2),(5i+2)) \leq 8(5i+2)$ for all integers $i \geq 0$. Thus, $\gamma_{dR}(P(5k,k)) \leq 8k$ for $k \equiv 2 \bmod 5$.

The statement for $k \equiv 3 \bmod 5$ follows from Table 5 by analogous argument. □

The next table provides a DRDF for $P(25,5)$. It is obtained from Table 4 by deleting columns 3 and 4 and altering only one entry in the original column 5, see Table 6.

Table 6. A DRDF of U_i that implies $\gamma_{dR}(P(25,5)) \leq 42$.

2	0	2	2	0	2	...	
2	0	0	2	0	0	...	
0	2	0	2	2	0	...	
0	2	0	0	2	0	...	
0	0	2	0	0	2	...	
0	1	2	3,4,5	6	7	...	merged columns
0	1	2	3	4	5	...	columns renamed

It is straightforward to check that the table provides a DRDF. Observe that we can in this way delete two columns and alter one column to obtain a DRDF of $\gamma_{dR}(P(5(5i), 5i)) \leq 8(5i) + 2$ from $\gamma_{dR}(P(5(5i+2), 5i+2)) \leq 8(5i+2)$ for all integers $i \geq 0$.

The same idea, applied to Table 5 gives a DRDF of $P(30, 6)$ of weight 50. We omit the details. Using the periodicity of the basic pattern, we have a construction that gives RDF showing $\gamma_{dR}(P(5(5i+1), 5i+1)) \leq 8(5i+1) + 2$ from $\gamma_{dR}(P(5(5i+3), (5i+3))) \leq 8(5i+3)$ for all integers $i \geq 0$.

Similarly, inserting two columns in the pattern comes with additional cost of $8 + 8 + 2 = 18$ legions, thus increasing the total weight by 18. For example, see Table 7.

Table 7. An alternative DRDF of U_i that shows $\gamma_{dR}(P(25,5)) \leq 42$ and $\gamma_{dR}(P(50,10)) \leq 82$.

2	0	0	2	0	2	0	2	0	0	2	...	
2	0	2	0	2	0	0	2	0	2	0	...	
0	0	2	0	2	0	2	0	0	2	0	...	
0	2	0	2	0	0	2	0	2	0	0	...	
0	2	0	2	0	2	0	0	2	0	2	...	
0	1	2	3'	2'	3	4	5	6	7	8	...	
0	1	2	3	4	5	6	7	8	9	10		columns renamed

For completeness, in Table 8 we give a RDF of $P(20,4)$ proving that $\gamma_{dR}(P(20,4)) \leq 34$. The construction starts with RDF of weight 16 for $\gamma_{dR}(P(10,2))$, and inserts two columns as in Table 7. In more detail, note that column 4 is a copy of column 2 and column 3 is a copy of column 1. Then, additional two legions are assigned to vertex u_{17} in column 1. It follows that $\gamma_{dR}(P(5(5i+4), (5i+4))) \leq 8(5i+4) + 2$ for all integers $i \geq 0$.

Table 8. A DRDF of U_i that implies $\gamma_{dR}(P(20,4)) \leq 34$.

2	0	2	0	2	
2	0	0	0	0	
0	2	0	2	0	
0	2	0	2	0	
0	2	2	0	2	
0	1'	2'	1	2	
0	1	2	3	4	columns renamed

Summarizing the arguments, we have a proof of the next proposition.

Proposition 6. *Let $k \equiv 0, 1, 4 \mod 5$. Then $\gamma_{dR}(P(5k, k)) \leq 8k + 2$.*

3.3. Double Roman Domination in $P(5k, k)$—Lower Bound

The proof of lower bound in Theorem 8 is based on several technical lemmas. In all proofs below we assume that f is a DRDF and there are no vertices with $f(v) = 1$. As before, let $U_i = \{u_i, u_{i+k}, u_{i+2k}, u_{i+3k}, u_{i+4k}\}$ and $V_i = \{v_i, v_{i+k}, v_{i+2k}, v_{i+3k}, v_{i+4k}\}$. Let us denote with W_i the weight of $H_i = V_i \cup U_i$, $W_i = f(H_i) = f(V_i \cup U_i)$.

Lemma 1. *Let f be DRDF f. Then*

if $f(U_i) = 0$ then $f(V_i) \geq 6$ and $W_i \geq 6$,
if $f(U_i) = 2$ then $f(V_i) \geq 5$ and $W_i \geq 7$,
if $f(U_i) = 3$ then $f(V_i) \geq 5$ and $W_i \geq 8$,
if $f(U_i) = 4$ then $f(V_i) \geq 4$ and $W_i \geq 8$,
if $f(U_i) = 5$ then $f(V_i) \geq 4$ and $W_i \geq 9$,
if $f(U_i) = 6$ then $f(V_i) \geq 3$ and $W_i \geq 9$,
if $f(U_i) = 7$ then $f(V_i) \geq 4$ and $W_i \geq 11$,
if $f(U_i) = 8$ then $f(V_i) \geq 3$ and $W_i \geq 11$, and
if $f(U_i) \geq 9$ then $W_i \geq 12$.

Proof. We will list all possible examples (up to the isomorphism), using the following notation:

$$\begin{pmatrix} f(u_i) & f(u_{i+k}) & f(u_{i+2k}) & f(u_{i+3k}) & f(u_{i+4k}) \\ f(v_i) & f(v_{i+k}) & f(v_{i+2k}) & f(v_{i+3k}) & f(v_{i+4k}) \end{pmatrix}.$$

- Case $f(U_i) \geq 9$. First, assume $f(U_i) = 9$. Excluding weights 1, the sum 9 can be achieved as $9 = 3 + 3 + 3$ or $9 = 3 + 2 + 2 + 2$. In the first case, three vertices among five can be chosen in two ways, either the two zeros are at adjacent columns or not. Similarly, in the second case, the 0 can either be next to 3, or not. Thus, we have 4 cases listed below. The values on V_i are chosen so that the total weight is minimal.

$$\begin{pmatrix} 3 & 3 & 3 & 0 & 0 \\ 0 & 0 & 0 & 3 & 0 \end{pmatrix}, \begin{pmatrix} 3 & 3 & 0 & 3 & 0 \\ 0 & 0 & 0 & 3 & 0 \end{pmatrix}, \begin{pmatrix} 3 & 2 & 2 & 2 & 0 \\ 0 & 0 & 2 & 0 & 2 \end{pmatrix}, \begin{pmatrix} 3 & 2 & 2 & 0 & 2 \\ 0 & 1 & 0 & 2 & 0 \end{pmatrix}.$$

In all cases we have $f(V_i) \geq 3$, thus $W_i \geq 9 + 3 = 12$.
Furthermore, if $f(U_i) = 10$ or $f(U_i) = 11$ then observe that $f(V_i) \geq 2$, and hence $W_i \geq 12$.

- Case $f(U_i) = 8$. Possible subcases with $8 = 3 + 3 + 2 = 2 + 2 + 2 + 2$ are

$$\begin{pmatrix} 3 & 3 & 2 & 0 & 0 \\ 0 & 0 & 0 & 3 & 0 \end{pmatrix}, \begin{pmatrix} 3 & 3 & 0 & 2 & 0 \\ 0 & 0 & 0 & 3 & 0 \end{pmatrix}, \begin{pmatrix} 3 & 2 & 3 & 0 & 0 \\ 0 & 0 & 2 & 0 & 2 \end{pmatrix}, \begin{pmatrix} 3 & 2 & 0 & 3 & 0 \\ 0 & 0 & 2 & 0 & 2 \end{pmatrix}$$

and

$$\begin{pmatrix} 2 & 2 & 2 & 2 & 0 \\ 0 & 2 & 0 & 0 & 2 \end{pmatrix}.$$

In all cases, $f(V_i) \geq 3$, thus $W_i \geq 8 + 3 = 11$.

- Case $f(U_i) = 7$. There is only one possibility, $7 = 3 + 2 + 2$, and we have the following subcases:

$$\begin{pmatrix} 3 & 2 & 2 & 0 & 0 \\ 0 & 0 & 2 & 0 & 2 \end{pmatrix}, \begin{pmatrix} 3 & 2 & 0 & 2 & 0 \\ 0 & 0 & 2 & 0 & 2 \end{pmatrix}, \begin{pmatrix} 3 & 2 & 0 & 0 & 2 \\ 0 & 0 & 2 & 2 & 0 \end{pmatrix}, \begin{pmatrix} 3 & 0 & 2 & 2 & 0 \\ 0 & 2 & 0 & 0 & 2 \end{pmatrix}.$$

It is obvious that $f(V_i) \geq 4$ and $W_i \geq 7 + 4 = 11$.

- Case $f(U_i) = 6$. This sum can be achieved as $6 = 3 + 3 = 2 + 2 + 2$. There are four subcases:

$$\begin{pmatrix} 3 & 3 & 0 & 0 & 0 \\ 0 & 0 & 0 & 3 & 0 \end{pmatrix}, \begin{pmatrix} 3 & 0 & 3 & 0 & 0 \\ 0 & 2 & 0 & 0 & 3 \end{pmatrix}, \begin{pmatrix} 2 & 2 & 2 & 0 & 0 \\ 0 & 0 & 2 & 0 & 2 \end{pmatrix}, \begin{pmatrix} 2 & 2 & 0 & 2 & 0 \\ 0 & 0 & 2 & 0 & 2 \end{pmatrix}.$$

In all cases the value $f(V_i)$ is at least 3, which implies $W_i \geq 6 + 3 = 9$.
- Case $f(U_i) = 5$. We have $5 = 3 + 2$, and two possibilities.

$$\begin{pmatrix} 3 & 2 & 0 & 0 & 0 \\ 0 & 0 & 2 & 0 & 2 \end{pmatrix}, \begin{pmatrix} 3 & 0 & 2 & 0 & 0 \\ 0 & 2 & 0 & 3 & 0 \end{pmatrix}.$$

Clearly, in both cases $f(V_i)$ must be at least 4, which implies $W_i \geq 5 + 4 = 9$.
- Case $f(U_i) = 4$. As $4 = 2 + 2$, we have two cases:

$$\begin{pmatrix} 2 & 2 & 0 & 0 & 0 \\ 0 & 0 & 2 & 0 & 2 \end{pmatrix}, \begin{pmatrix} 2 & 0 & 2 & 0 & 0 \\ 0 & 2 & 0 & 3 & 0 \end{pmatrix}.$$

As $f(V_i) \geq 4$ in both cases, we have $W_i \geq 4 + 4 = 8$.
- Case $f(U_i) = 3$. There is only one possible subcase $\begin{pmatrix} 3 & 0 & 0 & 0 & 0 \\ 0 & 0 & 3 & 0 & 2 \end{pmatrix}$ with $f(V_i) = 5$, thus $W_i \geq 3 + 5 = 8$.
- Case $f(U_i) = 2$. The only possible subcase is $\begin{pmatrix} 2 & 0 & 0 & 0 & 0 \\ 0 & 2 & 0 & 3 & 0 \end{pmatrix}$ with $f(V_i) = 5$, thus $W_i \geq 2 + 5 = 7$.
- Case $f(U_i) = 0$ has two possible subcases,

$$\begin{pmatrix} 0 & 0 & 0 & 0 & 0 \\ 0 & 3 & 0 & 0 & 3 \end{pmatrix}, \begin{pmatrix} 0 & 0 & 0 & 0 & 0 \\ 2 & 2 & 0 & 2 & 0 \end{pmatrix},$$

with $f(V_i) = 6$, thus $W_i \geq 0 + 6 = 6$.

This concludes the proof of lemma. □

In order to prove the lower bound in Theorem 8, we will need to consider the H_i with $W_i < 8$, thus by Lemma 1, the cases $W_i = 7$ ($f(U_i) = 2$, and $f(U_i) = 0$) or $W_i = 6$ ($f(U_i) = 0$). In the Figure 2 below all cases (up to the isomorphism) with $W_i = 6$ and $W_i = 7$ are drawn.

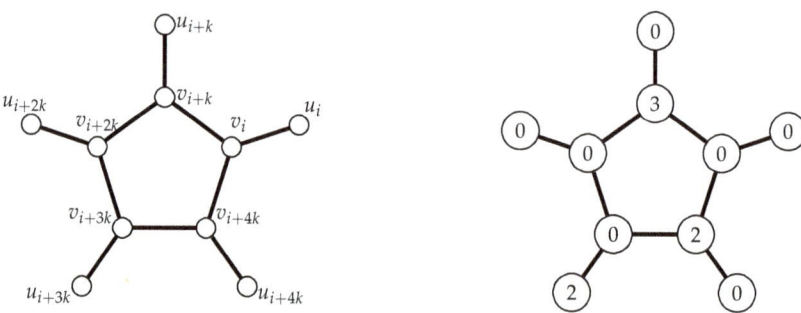

Figure 2. The standard drawing of H_i (**left**) and the case $f(H_i) = W_i = 7$ with $f(U_i) = 2$ (**right**).

First we consider the cases where two adjacent H_i have weights less than 8. Note that the proof of the next lemma also implies that it is not possible to have a DRDF with more that two consecutive $W_i < 8$.

Lemma 2.
(a) If $W_i = 6$ and $W_{i+1} = 6$ then $W_{i-1} \geq 12$ and $W_{i+2} \geq 12$.
(b) If $W_i = 7$ and $W_{i+1} = 7$ then $W_{i-1} \geq 11$ and $W_{i+2} \geq 11$.
(c) If $W_i = 6$ and $W_{i+1} = 7$ then $W_{i-1} \geq 11$, $W_{i+2} \geq 11$, and $W_{i-1} + W_{i+2} \geq 23$.

Proof. The proof will be derived in several steps using the notation introduced in Table 3. We only give the values on the outer cycle, and in addition, in some cases (for sets H_i and H_{i+1}) the values on the inner cycles are provided in parentheses, as $\mathbf{f(u_j)}(f(v_j))$. For other neighbor sets H_*, we will assume that the inner cycles V_* are completed such that the whole assignment is a DRD function. The weights W_i are estimated using the results of Lemma 1.

(a) Case $W_i = 6$ and $W_{i+1} = 6$ obviously implies that $f(U_i) = 0$ and $f(U_{i+1}) = 0$. There are two cases (see Figure 3), for which Table 9 (columns A1 and A2) show the minimal demands that the two neighboring vertices in $U_{i-1} \cup U_{i+1}$ have to fulfil. Without loss of generality, consider first H_i. Since $f(U_{i+1}) = 0$, we read from Table 9 that at least three vertices of U_{i-1} must have weights 3, thus $f(U_{i-1}) \geq 9$ and $W_{i-1} \geq 9 + 3 = 12$. By analogous reasoning, $W_{i+2} \geq 9 + 3 = 12$.

(b) Case $W_i = 7$ and $W_{i+1} = 7$ (see Figure 4). First, consider the case when $f(U_{i+1}) = 0$. Then, from Table 9 (columns A3 and A4) there are at least two vertices of U_{i-1} which must have weights 3, and two more vertices with weights at least two, thus $f(U_{i-1}) \geq 10$ and $W_{i-1} \geq 12$. By symmetry, $f(U_i) = 0$ implies $W_{i+2} \geq 12$.

Therefore, we may assume that $f(U_i) = 2$ and $f(U_{i+1}) = 2$. The DRDF for H_i is in Figure 2 (right). Considering neighbor sets H_{i-1} and H_{i+2} we have all subcases listed in Tables 10 and 11.

In Tables 10 and 11, we fix DRDF on H_i (second column), and consider all possible DRDF with $W_{i+1} = 7$ and $f(U_{i+1}) = 2$ (third column). The first and fourth columns provide the minimal f values in H_{i-1} and H_{i+2}, respectively. The labeled graph H_{i+1} in this case has no symmetries, hence we have to consider five rotations, and in each case two cases due to reflexion. Thus, we have ten cases in total, b1 to b10, outlined in Tables 10 and 11.

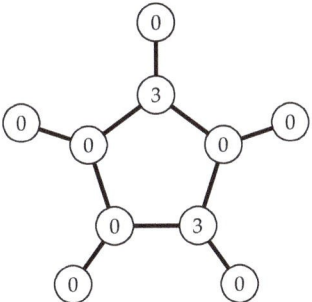

Figure 3. The two cases with $f(H_i) = W_i = 6$.

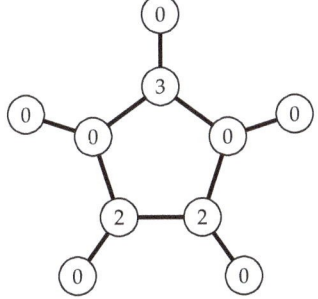
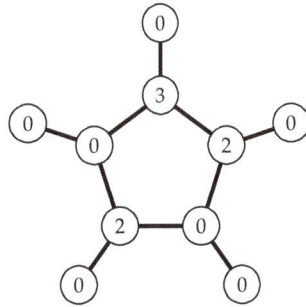

Figure 4. The two cases with $f(H_i) = W_i = 7$ and $f(U_i) = 0$.

Table 9. Demands for $U_{i-1} \cup U_{i+1}$ when $W_i = 6$ or $W_i = 7$ and $f(U_i) = 0$.

(A1)		(A2)		(A3)		(A4)	
0(0)	3+0/2+2	0(2)	0+2	0(0)	3+0/2+2	0(0)	3+0/2+2
0(3)	0	0(2)	0+2	0(3)	0	0(3)	0
0(0)	3+0/2+2	0(0)	3+0/2+2	0(0)	3+0/2+2	0(2)	2+0
0(0)	3+0/2+2	0(2)	0+2	0(2)	2+0	0(0)	3+0/2+2
0(3)	0	0(0)	3+0/2+2	0(2)	2+0	0(2)	2+0

Table 10. Subcases of $f(U_{i-1})$, $f(U_i)$, $f(U_{i+1})$ and $f(U_{i+2})$ with $W_i = 7$, $W_{i+1} = 7$ and $f(U_{i+1}) = 2$ (first part).

(b1)				(b2)				(b3)				(b4)				(b5)			
0	2(0)	2(0)	0	0	2(0)	0(0)	2	0	2(0)	0(3)	0	0	2(0)	0(0)	2	0	2(0)	0(2)	0
2	0(2)	0(2)	2	0	0(2)	2(0)	0	2	0(2)	0(0)	3	2	0(2)	0(3)	0	2	0(2)	0(0)	3
3	0(0)	0(0)	3	3	0(0)	0(2)	2	2	0(0)	2(0)	0	3	0(0)	0(0)	3	3	0(0)	0(3)	0
0	0(3)	0(3)	0	0	0(3)	0(0)	3	0	0(3)	0(2)	2	0	0(3)	2(0)	0	0	0(3)	0(0)	3
3	0(0)	0(0)	3	3	0(0)	0(3)	0	3	0(0)	0(0)	3	3	0(0)	0(2)	2	3	0(0)	2(0)	0

Table 11. Subcases of $f(U_{i-1})$, $f(U_i)$, $f(U_{i+1})$ and $f(U_{i+2})$ with $W_i = 7$, $W_{i+1} = 7$ and $f(U_{i+1}) = 2$ (second part).

(b6)				(b7)				(b8)				(b9)				(b10)			
0	2(0)	0(2)	2	0	2(0)	0(0)	2	0	2(0)	0(3)	0	0	2(0)	0(0)	2	0	2(0)	2(0)	0
2	0(2)	2(0)	0	0	0(2)	0(2)	2	2	0(2)	0(0)	3	2	0(2)	0(3)	0	2	0(2)	0(0)	3
3	0(0)	0(0)	3	3	0(0)	2(0)	0	2	0(0)	0(2)	2	3	0(0)	0(0)	3	3	0(0)	0(3)	0
0	0(3)	0(3)	0	0	0(3)	0(0)	3	0	0(3)	2(0)	0	0	0(3)	0(2)	2	0	0(3)	0(0)	3
3	0(0)	0(0)	3	3	0(0)	0(3)	0	3	0(0)	0(0)	3	3	0(0)	2(0)	0	3	0(0)	0(2)	2

From Tables 10 and 11 using the results of Lemma 1 we can estimate the weights W_i: **(b1)** $W_{i-1} \geq 8 + 4 = 12$, $W_{i+2} \geq 8 + 4 = 12$, **(b2)** $W_{i-1} \geq 6 + 5 = 11$, $W_{i+2} \geq 7 + 4 = 11$, **(b3)** $W_{i-1} \geq 7 + 4 = 11$, $W_{i+2} \geq 8 + 4 = 12$, **(b4)** $W_{i-1} \geq 8 + 4 = 12$, $W_{i+2} \geq 7 + 4 = 11$, **(b5)** $W_{i-1} \geq 8 + 4 = 12$, $W_{i+2} \geq 6 + 5 = 11$, **(b6)** $W_{i-1} \geq 8 + 4 = 12$, $W_{i+2} \geq 8 + 4 = 12$, **(b7)** $W_{i-1} \geq 6 + 5 = 11$, $W_{i+2} \geq 7 + 4 = 11$, **(b8)** $W_{i-1} \geq 7 + 4 = 11$, $W_{i+2} \geq 8 + 4 = 12$, **(b9)** $W_{i-1} \geq 8 + 4 = 12$, $W_{i+2} \geq 7 + 4 = 11$, **(b10)** $W_{i-1} \geq 8 + 4 = 12$, $W_{i+2} \geq 8 + 4 = 12$.

(c) Case $W_i = 6$ and $W_{i+1} = 7$. First, observe that in the case when $f(U_{i+1}) = 0$, the reasoning in case (a) and (b) implies that $W_{i-1} \geq 12$ and $W_{i+2} \geq 11$. So we can assume that $f(U_{i+1}) = 2$. As seen in Figure 3, three vertices in V_i could have weights 2 or two of them have weights 3. As we know, one vertex in U_{i+1} has weight 3 and the other one (two steps further) has weight 2, see Figure 2. We thus fix the assignment in H_i (second column), add all possible assignments in H_{i+1} (third column) and write the minimal weights on H_{i-1} and H_{i+2} in the first and fourth column. As each of the two assignments of H_i is reflexion symmetric, it is clear that there are exactly 10 different cases. All possible outcomes for sets $H_{i-1} \cup H_i \cup H_{i+1}$ are given below, Tables 12 and 13.

In Table 12, we find all subcases (c1 to c5) when two vertices in V_i have weights 3.

Table 12. First five subcases of $f(U_{i-1})$, $f(U_i)$, $f(U_{i+1})$ and $f(U_{i+2})$ with $W_i = 6$ and $W_{i+1} = 7$.

(c1)				(c2)				(c3)				(c4)				(c5)			
2	0(0)	2(0)	0	3	0(0)	0(2)	2	3	0(0)	0(0)	3	3	0(0)	0(3)	0	3	0(0)	0(0)	3
0	0(3)	0(0)	3	0	0(3)	2(0)	0	0	0(3)	0(2)	2	0	0(3)	0(0)	3	0	0(3)	0(3)	0
3	0(0)	0(3)	0	3	0(0)	0(0)	3	2	0(0)	2(0)	0	3	0(0)	0(2)	2	3	0(0)	0(0)	3
3	0(0)	0(0)	3	3	0(0)	0(3)	0	3	0(0)	0(0)	3	2	0(0)	2(0)	0	3	0(0)	0(2)	2
0	0(3)	0(2)	2	0	0(3)	0(0)	3	0	0(3)	0(3)	0	0	0(3)	0(0)	3	0	0(3)	2(0)	0

In Table 13, we have all subcases (c6 to c10) when three vertices in V_i have weights 2.

Table 13. Second five subcases of $f(U_{i-1})$, $f(U_i)$, $f(U_{i+1})$ and $f(U_{i+2})$ with $W_i = 6$ and $W_{i+1} = 7$.

	(c6)				(c7)				(c8)				(c9)				(c10)		
0	0(2)	2(0)	0	2	0(2)	0(2)	2	2	0(2)	0(0)	3	2	0(2)	0(3)	0	2	0(2)	0(0)	3
2	0(2)	0(0)	3	0	0(2)	2(0)	0	2	0(2)	0(2)	2	2	0(2)	0(0)	3	2	0(2)	0(3)	0
3	0(0)	0(3)	0	3	0(0)	0(0)	3	2	0(0)	2(0)	0	3	0(0)	0(2)	2	3	0(0)	0(0)	3
2	0(2)	0(0)	3	2	0(2)	0(3)	0	2	0(2)	0(0)	3	0	0(2)	2(0)	0	2	0(2)	0(2)	2
3	0(0)	0(2)	2	3	0(0)	0(0)	3	3	0(0)	0(3)	0	3	0(0)	0(0)	3	2	0(0)	2(0)	0

Similarly, as in case (b), we can estimate the weights W_i from Tables 12 and 13 using the results of Lemma 1: **(c1)** $W_{i-1} \geq 8 + 3 = 11$, $W_{i+2} \geq 8 + 4 = 12$, **(c2)** $W_{i-1} \geq 9 + 3 = 12$, $W_{i+2} \geq 8 + 4 = 12$, **(c3)** $W_{i-1} \geq 8 + 4 = 12$, $W_{i+2} \geq 8 + 4 = 12$, **(c4)** $W_{i-1} \geq 8 + 4 = 12$, $W_{i+2} \geq 8 + 4 = 12$, **(c5)** $W_{i-1} \geq 9 + 3 = 12$, $W_{i+2} \geq 8 + 4 = 12$, **(c6)** $W_{i-1} \geq 10 + 4 = 14$, $W_{i+2} \geq 8 + 4 = 12$, **(c7)** $W_{i-1} \geq 10 + 4 = 14$, $W_{i+2} \geq 8 + 4 = 12$, **(c8)** $W_{i-1} \geq 11 + 4 = 15$, $W_{i+2} \geq 8 + 4 = 12$, **(c9)** $W_{i-1} \geq 10 + 4 = 14$, $W_{i+2} \geq 8 + 4 = 12$, **(c10)** $W_{i-1} \geq 11 + 4 = 15$, $W_{i+2} \geq 8 + 4 = 12$. □

Lemma 3.

(a) If $W_i = 6$ and $W_{i-1} \geq 8$, $W_{i+1} \geq 8$ then either $(W_{i-1} + W_{i+1} \geq 20)$ or $(W_{i-2} + W_{i-1} \geq 19$, $W_{i+1} + W_{i+2} \geq 19$, and $W_{i-2} + W_{i-1} + W_{i+1} + W_{i+2} \geq 39)$.

(b) If $W_i = 7$ and $W_{i-1} \geq 8$, $W_{i+1} \geq 8$ then either $(W_{i-1} + W_{i+1} \geq 18)$ or $(W_{i-2} + W_{i-1} \geq 18$ and $W_{i+1} + W_{i+2} \geq 18)$.

Proof. As in the proof of Lemma 2, we will use the notation introduced in Table 3. We give only the values on the outer cycle, and in some cases the values on the inner cycles are provided in parenthesis, as $\mathbf{f}(\mathbf{u_j})(f(v_j))$. For other neighbor sets H_* we will assume that the inner cycles V_* are completed such that the whole assignment is a DRD function.

(a) Case $W_i = 6$ implies $f(U_i) = 0$ (Figure 3). Recall that either three vertices in V_i have weight 2 or two of them have weight 3, and that Table 9 gives the demands that need to be fulfilled by the neighboring H_*. As $W_{i-1} \geq 8$ and $W_{i+1} \geq 8$, according to lemma 1, we have $f(U_{i-1}) \geq 3$ and $f(U_{i+1}) \geq 3$. We may also assume that $f(U_{i-1}) \leq f(U_{i+1})$. Thus, we have $f(U_{i-1}) \in \{3, 4, 5, 6\}$, and all cases are analyzed in Tables 14–16. Note that there is only one case for $f(U_{i-1}) = 6 = 2 + 2 + 2$ (and $f(U_{i+1}) = 6 = 2 + 2 + 2$), because if $f(U_{i-1}) = 6 = 3 + 3$ then $f(U_{i+1}) = 3 < 6$.

Table 14. Subcases of $f(U_{i-1})$, $f(U_i)$ and $f(U_{i+1})$ with $W_i = 6 = 3 + 3$.

	(A1-1)*			(A1-2)*			(A1-3)*			(A1-4)*	
3	0(0)	0	0	0(0)	3	0	0(0)	3	0	0(0)	3
0	0(3)	0	0	0(3)	0	0	0(3)	0	0	0(3)	0
0	0(0)	3	3	0(0)	0	3	0(0)	0	2	0(0)	2
0	0(0)	3	0	0(0)	3	2	0(0)	2	2	0(0)	2
0	0(3)	0	0	0(3)	0	0	0(3)	0	0	0(3)	0
	(A1-5)*			(A1-6)			(A1-7)			(A1-8)	
3	0(0)	0	2	0(0)	2	2	0(0)	2	2	0(0)	2
0	0(3)	0	0	0(3)	0	0	0(3)	0	0	0(3)	0
2	0(0)	2	3	0(0)	0	2	0(0)	2	2	0(0)	2
0	0(0)	3	0	0(0)	3	0	0(0)	3	2	0(0)	2
0	0(3)	0	0	0(3)	0	0	0(3)	0	0	0(3)	0

Reading Table 14 we observe that weights are **(A1-1)*** $W_{i-1} \geq 3 + 5 = 8$, $W_{i+1} \geq 6 + 3 = 9$, **(A1-2)*** $W_{i-1} \geq 3 + 5 = 8$, $W_{i+1} \geq 6 + 5 = 11$, **(A1-3)*** $W_{i-1} \geq 5 + 4 = 9$, $W_{i+1} \geq 5 + 5 = 10$, **(A1-4)*** $W_{i-1} \geq 4 + 4 = 8$, $W_{i+1} \geq 7 + 4 = 11$, **(A1-5)*** $W_{i-1} \geq 5 + 5 = 10$, $W_{i+1} \geq 5 + 4 = 9$, **(A1-6)** $W_{i-1} \geq 5 + 5 = 10$, $W_{i+1} \geq 5 + 5 = 10$, **(A1-7)** $W_{i-1} \geq 4 + 5 = 9$, $W_{i+1} \geq 7 + 4 = 11$, **(A1-8)** $W_{i-1} \geq 6 + 4 = 10$, $W_{i+1} \geq 6 + 4 = 10$, so in cases (A1-6), (A1-7), and (A1-8), $W_{i-1} + W_{i+1} \geq 20$. However, in the first five cases (labelled with asterisk), $W_{i-1} + W_{i+1} < 20$, and therefore we need to consider H_{i-2} and H_{i+2} to conclude the proof of assertion (a) of Lemma 3, see Table 15.

Table 15. Subcases* of $f(U_{i-2}), f(U_{i-1}), f(U_i), f(U_{i+1})$ and $f(U_{i+2})$ with $W_{i-1} + W_{i+1} < 20$.

(A1-1)*					(A1-2)*					(A1-3)*				
0	3(0)	**0**(0)	0(3)	0	3	0(0)	**0**(0)	3(0)	0	3	0(0)	**0**(0)	3(0)	0
3	0(0)	**0**(3)	0(0)	3	2	0(2)	**0**(3)	0(0)	3	2	0(2)	**0**(3)	0(0)	3
0	0(3)	**0**(0)	3(0)	0	0	3(0)	**0**(0)	0(3)	0	0	3(0)	**0**(0)	0(3)	0
3	0(0)	**0**(0)	3(0)	0	3	0(0)	**0**(0)	3(0)	0	0	2(0)	**0**(0)	2(0)	0
2	0(2)	**0**(3)	0(0)	3	0	0(3)	**0**(3)	0(2)	2	2	0(2)	**0**(3)	0(2)	2
(A1-4)*					(A1-5)*									
3	0(0)	**0**(0)	3(0)	0	0	3(0)	**0**(0)	0(0)	3					
2	0(2)	**0**(3)	0(2)	2	2	0(2)	**0**(3)	0(2)	2					
0	2(0)	**0**(0)	2(0)	0	0	2(0)	**0**(0)	2(0)	0					
0	2(0)	**0**(0)	2(0)	0	0	0(3)	**0**(0)	3(0)	0					
2	0(2)	**0**(3)	0(2)	2	3	0(0)	**0**(3)	0(2)	2					

From Table 15 we can estimate weights **(A1-1)*** $W_{i-2} \geq 8 + 4 = 12$, $W_{i-1} \geq 3 + 5 = 8$, $W_{i+1} \geq 6 + 3 = 9$, $W_{i+2} \geq 6 + 5 = 11$, **(A1-2)*** $W_{i-2} \geq 8 + 4 = 12$, $W_{i-1} \geq 3 + 5 = 8$, $W_{i+1} \geq 6 + 5 = 11$, $W_{i+2} \geq 5 + 5 = 10$, **(A1-3)*** $W_{i-2} \geq 7 + 4 = 11$, $W_{i-1} \geq 5 + 4 = 9$, $W_{i+1} \geq 5 + 5 = 10$, $W_{i+2} \geq 5 + 5 = 10$, **(A1-4)*** $W_{i-2} \geq 7 + 4 = 11$, $W_{i-1} \geq 4 + 4 = 8$, $W_{i+1} \geq 7 + 4 = 11$, $W_{i+2} \geq 4 + 5 = 9$, **(A1-5)*** $W_{i-2} \geq 5 + 5 = 10$, $W_{i-1} \geq 5 + 5 = 10$, $W_{i+1} \geq 5 + 4 = 9$, $W_{i+2} \geq 7 + 4 = 11$. Here, in all cases, $W_{i-2} + W_{i-1} \geq 19$, $W_{i+1} + W_{i+2} \geq 19$ and $W_{i-2} + W_{i-1} + W_{i+1} + W_{i+2} \geq 39$.

Table 16. Subcases of $f(U_{i-1}), f(U_i)$ and $f(U_{i+1})$ with $W_i = 6 = 2 + 2 + 2$.

(A2-1)			(A2-2)			(A2-3)			(A2-4)			(A2-5)		
0	0(2)	2	0	0(2)	2	2	0(2)	0	0	0(2)	2	0	0(2)	2
0	0(2)	2	0	0(2)	2	0	0(2)	2	0	0(2)	2	0	0(2)	2
3	0(0)	0	2	0(0)	2	2	0(0)	2	2	0(0)	2	2	0(0)	2
0	0(2)	2	2	0(2)	0	0	0(2)	2	0	0(2)	2	0	0(2)	2
3	0(0)	0	2	0(0)	2	2	0(0)	2	0	0(0)	2	3	0(0)	0
(A2-6)			(A2-7)			(A2-8)			(A2-9)			(A2-10)		
0	0(2)	2	2	0(2)	0	0	0(2)	2	2	0(2)	0	2	0(2)	0
0	0(2)	2	0	0(2)	2	0	0(2)	2	0	0(2)	2	2	0(2)	0
3	0(0)	0	3	0(0)	0	3	0(0)	0	2	0(0)	2	2	0(0)	2
2	0(2)	0	0	0(2)	2	0	0(2)	2	2	0(2)	0	0	0(2)	2
0	0(0)	3	0	0(0)	3	0	0(0)	3	0	0(0)	3	0	0(0)	3
(A2-11)			(A2-12)			(A2-13)			(A2-14)			(A2-15)		
0	0(2)	2	2	0(2)	0	2	0(2)	0	2	0(2)	0	2	0(2)	0
0	0(2)	2	2	0(2)	2	2	0(2)	0	0	0(2)	2	2	0(2)	0
2	0(0)	2	2	0(0)	2	0	0(0)	3	0	0(0)	3	0	0(0)	3
2	0(2)	0	0	0(2)	2	2	0(2)	0	2	0(2)	0	0	0(2)	2
0	0(0)	3	0	0(0)	3	0	0(0)	3	0	0(0)	3	0	0(0)	3

Reading Table 16 we observe that **(A2-1)** $W_{i-1} \geq 6 + 5 = 11$, $W_{i+1} \geq 6 + 4 = 10$, **(A2-2)** $W_{i-1} \geq 6 + 4 = 10$, $W_{i+1} \geq 8 + 4 = 12$, **(A2-3)** $W_{i-1} \geq 6 + 4 = 10$, $W_{i+1} \geq 8 + 4 = 12$, **(A2-4)** $W_{i-1} \geq 4 + 5 = 9$, $W_{i+1} \geq 10 + 4 = 14$, **(A2-5)** $W_{i-1} \geq 5 + 5 = 10$, $W_{i+1} \geq 8 + 4 = 12$, **(A2-6)** $W_{i-1} \geq 5 + 4 = 9$, $W_{i+1} \geq 7 + 4 = 11$, **(A2-7)** $W_{i-1} \geq 5 + 5 = 10$, $W_{i+1} \geq 7 + 4 = 11$, **(A2-8)** $W_{i-1} \geq 3 + 5 = 8$, $W_{i+1} \geq 9 + 3 = 12$, **(A2-9)** $W_{i-1} \geq 6 + 4 = 10$, $W_{i+1} \geq 7 + 4 = 11$, **(A2-10)** $W_{i-1} \geq 6 + 4 = 10$, $W_{i+1} \geq 7 + 4 = 11$, **(A2-11)** $W_{i-1} \geq 4 + 4 = 8$, $W_{i+1} \geq 9 + 4 = 13$, **(A2-12)** $W_{i-1} \geq 4 + 5 = 9$, $W_{i+1} \geq 9 + 4 = 13$, **(A2-13)** $W_{i-1} \geq 6 + 4 = 10$, $W_{i+1} \geq 6 + 5 = 11$, **(A2-14)** $W_{i-1} \geq 4 + 5 = 9$, $W_{i+1} \geq 8 + 4 = 12$, **(A2-15)** $W_{i-1} \geq 4 + 4 = 8$, $W_{i+1} \geq 8 + 4 = 12$, so in all cases $W_{i-1} + W_{i+1} \geq 20$, which proves the assertion (a) of Lemma 3.

(b) Case $W_i = 7$. First, assume that $f(U_i) = 2$ (see Figure 2), so one vertex in U_i has weight 3 and the other one has weight 2. Possible (due to symmetry) solutions for the whole set $H_{i-1} \cup H_i \cup H_{i+1}$ are considered in the following Table 17.

Table 17. Possible values for $f(U_{i-1})$, $f(U_i)$ and $f(U_{i+1})$ with $W_i = 7$ and $f(U_i) = 2$.

		(B1)		
-	0	2(0)	0	-
-	2/0	0(2)	0/2	-
-	0/3/2	0(0)	3/0/2	-
-	0	0(3)	0	-
-	0/3/2	0(0)	3/0/2	-
$i-2$	$i-1$	i	$i+1$	$i+2$

Without loss of generality, assume that $f(U_{i-1}) \leq f(U_{i+1})$. As $W_{i-1} \geq 8$ and $W_{i+1} \geq 8$ we have the following subcases (see Table 18).

Table 18. Subcases of $f(U_{i-1})$, $f(U_i)$ and $f(U_{i+1})$ with $W_i = 7$, $f(U_i) = 2$, $W_{i-1} \geq 8$ and $W_{i+1} \geq 8$.

	(B1-1)			(B1-2)			(B1-3)			(B1-4)*	
0	2(0)	0	0	2(0)	0	0	2(0)	0	0	2(0)	0
2	0(2)	0	0	0(2)	0	2	0(2)	0	0	0(2)	2
0	0(0)	3	0	0(0)	3	2	0(0)	2	0	0(0)	3
0	0(3)	0	0	0(3)	0	0	0(3)	0	0	0(3)	0
3	0(0)	0	2	0(0)	2	0	0(0)	3	3	0(0)	0
	(B1-5)			(B1-6)							
0	2(0)	0	0	2(0)	0						
0	0(2)	2	0	0(2)	2						
3	0(0)	0	2	0(0)	2						
0	0(3)	0	0	0(3)	0						
0	0(0)	3	2	0(0)	2						

Reading Table 18 we observe that in all cases except one (B1-4) we have $W_{i-1} + W_{i+1} \geq 18$. Indeed, in **(B1-1)** $W_{i-1} \geq 5 + 5 = 10$, $W_{i+1} \geq 3 + 5 = 8$, **(B1-2)** $W_{i-1} \geq 4 + 5 = 9$, $W_{i+1} \geq 5 + 5 = 10$, **(B1-3)** $W_{i-1} \geq 4 + 4 = 8$, $W_{i+1} \geq 5 + 5 = 10$, **(B1-4)*** $W_{i-1} \geq 3 + 5 = 8$, $W_{i+1} \geq 5 + 4 = 9$, **(B1-5)** $W_{i-1} \geq 3 + 5 = 8$, $W_{i+1} \geq 5 + 5 = 10$, **(B1-6)** $W_{i-1} \geq 4 + 5 = 9$, $W_{i+1} \geq 6 + 4 = 10$. In case **(B1-4)** we need to consider weights on H_{i-2} and H_{i+2} to conclude the proof of assertion (b) of Lemma 3.

As seen from Table 19, in case **(B1-4)** we can confirm $W_{i-2} + W_{i-1} \geq 18$ and $W_{i+1} + W_{i+2} \geq 18$.

Table 19. Subcase **(B1-4)** with $W_{i-2} \geq 7 + 4 = 11$, $W_{i-1} \geq 3 + 5 = 8$, $W_{i+1} \geq 5 + 4 = 9$ and $W_{i+2} \geq 5 + 4 = 9$.

		(B1-4)*		
2	0(0)	2(0)	0(2)	0
0	0(3)	0(2)	2(0)	0
3	0(0)	0(0)	3(0)	0
2	0(2)	0(3)	0(2)	2
0	3(0)	0(0)	0(0)	3

Finally, if $f(U_i) = 0$ (see Figure 4), then observe that in comparison to the case $f(U_i) = 2$, there must be at least one more vertex of weight 2 in $U_{i-1} \cup U_{i+1}$. We omit detailed analysis of the cases that confirm $W_{i-1} + W_{i+1} \geq 18$.
 □

Theorem 8. *For all k we have $\gamma_{dR}(P(5k, k)) \geq 8k$.*

Proof. We use the discharging method (see [14]). The basic idea is as follows. Assume that we have a DRDF. Consider certain subgraphs, in our case the subgraphs H_i, that are induced on $V_i \cup U_i$. Define some rules how the weights of heavy subgraphs are discharged to the neighbors such that the total weight does not change. Observe the weights of subgraphs after discharging.

In our case, the discharging rule is simple: If $f(H_i) = W_i > 8$ then H_i sends $\frac{1}{2}(W_i - 8)$ to H_{i-i} and to H_{i-i}. The new charge of H_i is thus 8. We denote the charges after the first round by W_i^*.

Now we show that if $f(H_i) < 8$ then, after at most four rounds of discharging, the new charge W_i^{****} of H_i is at least 8. Note that once the charge of H_i is at least 8, discharging will never decrease its charge below 8.

First, let us consider the cases of Lemma 2.

(a) If $W_i = 6$ and $W_{i+1} = 6$ then $W_{i-1} \geq 12$ and $W_{i+2} \geq 12$. After discharging, $W_i^* = 8$ and $W_{i+1}^* = 8$, as needed.

(b) If $W_i = 7$ and $W_{i+1} = 7$ then $W_{i-1} \geq 11$ and $W_{i+2} \geq 11$. After discharging we have $W_i^* = 7 + \frac{3}{2} \geq 8$ and $W_{i+1}^* = 7 + \frac{3}{2} \geq 8$.

(c) If $W_i = 6$ and $W_{i+1} = 7$ then $W_{i-1} \geq 11$, $W_{i+2} \geq 11$, and $W_{i-1} + W_{i+2} \geq 23$. Assume that $W_{i-1} \geq 11$, $W_{i+2} \geq 12$. After the first round of discharging, we get $W_i^* \geq 6 + \frac{3}{2} = 7 + \frac{1}{2}$ and $W_{i+1}^* \geq 7 + 2 = 9$. However, after the second round of discharging, we have $W_i^{**} \geq 7 + \frac{1}{2} + \frac{1}{2} = 8$.
If $W_{i-1} \geq 12$, $W_{i+2} \geq 11$, then observe that already $W_i^* \geq 8$ and $W_{i+1}^* \geq 8$.

Next, we consider the cases of Lemma 3.

(a) $W_i = 6$ and $W_{i-1} \geq 8$, $W_{i+1} \geq 8$. If $W_{i-1} + W_{i+1} \geq 20$, then $W_i^* \geq 6 + \frac{(20-16)}{2} = 8$, and we are done. Otherwise, by Lemma 3, $W_{i-2} + W_{i-1} \geq 19$, $W_{i+1} + W_{i+2} \geq 19$, and $W_{i-2} + W_{i-1} + W_{i+1} + W_{i+2} \geq 39$. Assume that $W_{i-2} + W_{i-1} = 19$ and distinguish two cases.

(a11) $W_{i-1} = 9$ and $W_{i-2} = 10$. In the first round of discharging, H_i receives $\frac{1}{2}$ from H_{i-1}, and $W_{i-1}^* = 9 - 2 \cdot \frac{1}{2} + 1 = 9$. In the second round of discharging, H_i again receives $\frac{1}{2}$ from H_{i-1}, and in total H_i receives charge 1 from the left side.

(a12) $W_{i-1} = 8$ and $W_{i-2} = 11$. After the first round of discharging, $W_{i-1}^* = 8 + \frac{3}{2}$. In the second round of discharging, H_{i-1} sends $\frac{3}{4}$ to its neighbors. Thus, H_i receives $\frac{3}{4}$ from the left side.

Recall that by Lemma 3, $W_{i-2} + W_{i-1} = 19$ implies $W_{i+1} + W_{i+2} \geq 20$, and distinguish two cases.

(a21) $W_{i+1} = 9$ and $W_{i+2} \geq 11$. In the first round of discharging, H_i receives $\frac{1}{2}$ from H_{i+1}, and $W_{i+1}^* = 9 - 2 \cdot \frac{1}{2} + \frac{3}{2} = 8 + \frac{3}{2}$. In the second round of discharging, H_i again receives $\frac{3}{4}$ from H_{i+1}, so in total H_i receives charge $\frac{5}{4}$ from the right side.

(a22) $W_{i+1} = 8$ and $W_{i+2} \geq 12$. After the first round of discharging, $W_{i+1}^* \geq 8 + 2 = 10$. In the second round of discharging, H_i receives charge 1 from H_{i+1}, and also $W_{i+2}^{**} \geq 8 + 1 = 9$. Thus, after the third round $W_{i+1}^{***} \geq 8 + \frac{1}{2}$, and, in the fourth round H_i receives charge $\frac{1}{4}$ from H_{i+1}. So, in total H_i receives charge $\frac{5}{4}$ from the right side.

Summarizing, H_i receives charge at least $\frac{3}{4}$ from the left side, and at least $\frac{5}{4}$ from the right side. Hence $W_i^{****} \geq 6 + 2 = 8$, as claimed.

(b) $W_i = 7$ and $W_{i-1} \geq 8$, $W_{i+1} \geq 8$. If $W_{i-1} + W_{i+1} \geq 18$ then $W_i^* \geq 7 + (18 - 16)\frac{1}{2} = 8$, as needed. Otherwise, $W_{i-2} + W_{i-1} \geq 18$ and $W_{i+1} + W_{i+2} \geq 18$. We now show that $W_{i-2} + W_{i-1} = 18$ implies that H_i will in two rounds receive at least $\frac{1}{2}$ charge from the left side. Consider two cases.

(b1) $W_{i-1} = 9$. In the first round of discharging, H_i receives $\frac{1}{2}$ from the left side.

(b2) If $W_{i-1} = 8$, then $W_{i-2} \geq 10$. After the first round of discharging, $W_{i-1}^* \geq 8 + 1$. In the second round of discharging, H_{i-1} sends at least $\frac{1}{2}$ to its neighbors.

Thus, H_i receives at least $\frac{1}{2}$ from the left side. By analogous reasoning, $W_{i+1} + W_{i+2} \geq 18$ implies that H_i receives at least $\frac{1}{2}$ from the right side. Consequently, in total H_i receives at least $\frac{1}{2} + \frac{1}{2}$, as claimed.

Therefore, after discharging, each subgraph H_i has weight W_i at least 8, and consequently, the total weight is at least $8k$, as claimed. □

4. Domination in Generalized Petersen Graphs $P(20k, k)$

In this section, we discuss how the constructions, and the corresponding upper bounds can be extended from $P(c_0 k, k)$ to $P((hc_0)k, k)$, for $h = 2, 3, \ldots$. In particular, we obtain improved upper bounds for $P(20k, k)$ from upper bounds for $P(4k, k)$ and $P(5k, k)$.

First, we recall the notion of covering graph and h-lift to observe that $P(20k, k)$ is a covering graph of both $P(4k, k)$ and $P(5k, k)$. For basic information on covering graphs see [39]. Here we follow the approach used in [40]. Let $G = (V_1, E_1)$ and $H = (V_2, E_2)$ be two graphs, and let $p : V_2 \to V_1$ be a surjection. We will call p a *covering map* from H to G if for each $v \in V_2$, the restriction of p to the neighborhood of $v \in V_2$ is a bijection onto the neighborhood of $p(v)$ in G. In other words, p maps edges incident to v one-to-one onto edges incident to $p(v)$. H is called a *covering graph*, or a lift, of G if there exists a covering map from H to G. Assuming H is a lift of G with a covering map p. If p has a property that for every vertex $v \in V(G)$, its *fiber* $p^{-1}(v)$ has exactly h elements, we call H a h-lift of G.

Obviously, a long cycle may be a covering graph of shorter cycles. For example, the cycle C_{100} is a 2-lift of C_{50}, considering the surjection $p(v_i) = v_{i \bmod 50}$. Furthermore, C_{100} is also a 25-lift of C_4, etc.

Proposition 7. *Let $k \geq 1$, $c_0 \geq 3$, and $h \geq 2$. Petersen graph $P((hc_0)k, k)$ is a h-lift of $P(c_0 k, k)$.*

Proof. Consider the surjection $p : V(P((hc_0)k, k)) \to V(P(c_0 k, k))$ defined by $p(v_i) = v_{i \bmod h}$, and $p(u_i) = u_{i \bmod h}$. □

Proposition 8. $\gamma_{dR}(P((hc_0)k, k)) \leq h \gamma_{dR}(P(c_0 k, k))$.

Proof. Let f be a DRDF of $P(c_0 k, k)$. Define $\tilde{f}$ as $\tilde{f}(v) = f(p(v))$ and observe that $\tilde{f}$ is a DRDF of $P((hc_0)k, k)$. Clearly, the weight $\tilde{f}(P((hc_0)k, k))$ is exactly $h \tilde{f}(P((c_0 k, k))$. We omit the details. □

Corollary 3. $\gamma_{dR}(P(20k, k)) \leq 5 \gamma_{dR}(P(4k, k)) \leq 30k + 15$.

As $\gamma_{dR}(P(20k, k)) \geq \frac{3}{2} 20k = 30k$ by Corollary 1, we also have

Corollary 4. *If $k \equiv 1 \bmod 2$, then $\gamma_{dR}(P(20k, k)) = 30k$.*

Applying Proposition 8 to the case $P(20k, k)$ and $P(5k, k)$, we obtain another Corollary.

Corollary 5. $\gamma_{dR}(P(20k, k)) \leq 4 \gamma_{dR}(P(5k, k)) = 32k + 8$.

Clearly, the upper bound in Corollary 5 is only better for $k = 1, 2$, and 3. In these cases, we obtain

$$\gamma_{dR}(P(20, 1)) \leq 40, \quad \gamma_{dR}(P(40, 2)) \leq 64, \quad \gamma_{dR}(P(60, 3)) \leq 96, \qquad (3)$$

which, together with Corollary 3 and Theorem 3 implies Theorem 7.

Note that this bound is a considerable improvement over the general bounds given in Theorem 4 [12]. Indeed, the upper bound (2) grows as $\mathcal{O}(30k)$. The bounds in Theorem 4 are of the from $\frac{3}{2}(20k)F(k) + \frac{5k}{4} + C$, where $\lim_{k \to \infty} F(k) = 1$, so the asymptotic growth is $\mathcal{O}(30k + \frac{5k}{4})$.

5. Conclusions and Future Work

In this paper, we have extended the known results on double Roman domination of families $P(ck, k)$ of generalized Petersen graphs, by adding either exact values or bounds with gap at most 2 for the family $P(5k, k)$. This naturally continues previous work, where the families $P(3k, k)$ and $P(4k, k)$ were studied.

There are several interesting related questions that open avenues for future work. For example:

- Find closed expressions or good lower and upper bounds for $\gamma_{dR}(P(6k, k))$. Which graphs among $P(6k, k)$ are double Roman?
- The method used here to improve bounds for $\gamma_{dR}(P(20k, k))$ using $\gamma_{dR}(P(4k, k))$ and $\gamma_{dR}(P(5k, k))$ may be used to improve bounds for $\gamma_{dR}(P(ck, k))$ for larger c.
- Can the small gaps between lower and upper bounds for $\gamma_{dR}(P(5k, k))$ (and, also for $\gamma_{dR}(P(4k, k))$) be resolved by finding and proving exact values?

The authors believe that this study has solved the problem on $P(5k, k)$ to the limits of the standard method. These methods may be sufficient to handle the problem, e.g., on $P(6k, k)$, but probably can not be applied to much larger c. Covering graphs, as indicated in Section 4, may provide a tool to provide improved bounds for larger c. On the other hand, the gaps between the lower and upper bounds in some cases may be solved by other methods, see for example [41] and the references there.

More generally, this work again shows the power of the discharging method. The discharging method is most well known for its central role in the proof of the four color theorem. This proof technique was extensively applied to study various graph coloring problems, in particular on planar graphs. In [14], it is shown that a suitably altered discharging technique can also be used on domination type problems and is illustrated on the double Roman domination on some generalized Petersen graphs. Here, we apply the method to another family of graphs and the same problem. This may encourage future applications to other domination type problems.

Author Contributions: D.R.P. and J.Ž. contributed equally to this work. All authors have read and agreed to the published version of the manuscript.

Funding: This work was supported in part by the Slovenian Research Agency ARRS (grants P2-0248, J2-2512, and J1-1693).

Institutional Review Board Statement: Not applicable.

Informed Consent Statement: Not applicable.

Data Availability Statement: Not applicable.

Acknowledgments: The authors wish to sincerely thank to the anonymous reviewers for their constructive comments.

Conflicts of Interest: The authors declare no conflict of interest.

References

1. Beeler, R.A.; Haynes, T.W.; Hedetniemi, S.T. Double Roman domination. *Discrete Appl. Math.* **2016**, *211*, 23–29. [CrossRef]
2. Cockayne, E.J.; Dreyer, P.A., Jr.; Hedetniemi, S.M.; Hedetniemi, S.T. Roman domination in graphs. *Discret. Math.* **2004**, *278*, 11–22. [CrossRef]
3. ReVelle, C.S.; Rosing, K.E. Defendens Imperium Romanum: A classical problem in military strategy. *Am. Math. Mon.* **2000**, *107*, 585–594. [CrossRef]
4. Stewart, I. Defend the Roman Empire! *Sci. Am.* **1999**, *281*, 136–138. [CrossRef]
5. Arquilla, J.; Fredricksen, H. "Graphing"—An Optimal Grand Strategy. *Mil. Oper. Res.* **1995**, *1*, 3–17. [CrossRef]
6. Hening, M.A. Defending the Roman Empire from Multiple Attacks. *Discret. Math.* **2003**, *271*, 101–115. [CrossRef]
7. Ahangar, H.A.; Chellali, M.; Sheikholeslami, S.M. On the double Roman domination in graphs. *Discret. Appl. Math.* **2017**, *232*, 1–7.
8. Barnejee, S.; Henning, M.A.; Pradhan, D. Algorithmic results on double Roman domination in graphs. *J. Comb. Optim.* **2020**, *39*, 90–114.
9. Poureidi, A.; Rad, N.J. On algorithmic complexity of double Roman domination. *Discret. Appl. Math.* **2020**, *285*, 539–551.
10. Zhang, X.; Li, Z.; Jiang, H.; Shao, Z. Double Roman domination in trees. *Inf. Process. Lett.* **2018**, *134*, 31–34. [CrossRef]

11. Poureidi, A. A linear algorithm for double Roman domination of proper interval graphs. *Discret. Math. Algorithms Appl.* **2020**, *12*, 2050011. [CrossRef]
12. Gao, H.; Huang, J.; Yang, Y. Double Roman Domination in Generalized Petersen Graphs. *Bull. Iran. Math. Soc.* **2021**, 1–10. [CrossRef]
13. Jiang, H.; Wu, P.; Shao, Z.; Rao, Y.; Liu, J. The double Roman domination numbers of generalized Petersen graphs $P(n,2)$. *Mathematics* **2018**, *6*, 206. [CrossRef]
14. Shao, Z.; Wu, P.; Jiang, H.; Li, Z.; Žerovnik, J.; Zhang, X. Discharging approach for double Roman domination in graphs. *IEEE Acces* **2018**, *6*, 63345–63351. [CrossRef]
15. Shao, Z.; Erveš, R.; Jiang, H.; Peperko, A.; Wu, P.; Žerovnik, J. Double Roman graphs in $P(3k,k)$. *Mathematics* **2021**, *9*, 336. [CrossRef]
16. Amjadi, J.; Nazari-Moghaddam, S.; Sheikholeslami, S.M.; Volkmann, L. An upper bound on the double Roman domination number. *J. Comb. Optim.* **2018**, *36*, 81–89. [CrossRef]
17. Maimani, H.R.; Momeni, M.; Mahid, F.R.; Sheikholeslami, S.M. Independent double Roman domination in graphs. *AKCE Int. J. Graphs Comb.* **2020**, *17*, 905–910. [CrossRef]
18. Mobaraky, B.P.; Sheikholeslami, S.M. Bounds on Roman domination numbers of graphs. *Mat. Vesnik* **2008**, *60*, 247–253.
19. Volkmann, L. Double Roman domination and domatic numbers of graphs. *Commun. Comb. Optim.* **2018**, *3*, 71–77.
20. Garey, M.R.; Johnson, D.S. *Computers and Intractability: A Guide to the Theory of NP-Completeness*; W. H. Freeman and Co.: San Francisco, CA, USA, 1979.
21. Haynes, H.W.; Hedetniemi, S.; Slater, P. *Fundamentals of Domination in Graphs*; Marcel Dekker: New York, NY, USA, 1998.
22. Haynes, H.W.; Hedetniemi, S.; Slater, P. *Domination in Graphs: Advanced Topics*; Marcel Dekker: New York, NY, USA, 1998.
23. Ore, O. *Theory of Graphs*; American Mathematical Society: Providence, RI, USA, 1967.
24. Henning, M.A. A Characterization of Roman trees. *Discuss. Math. Graph Theory* **2002**, *22*, 325–334. [CrossRef]
25. Liu, C.H.; Chang, G.J. Upper bounds on Roman domination numbers of graphs. *Discret. Math.* **2012**, *312*, 1386–1391. [CrossRef]
26. Liu, C.H.; Chang, G.J. Roman domination on 2-connected graphs. *SIAM J. Discret. Math.* **2012**, *26*, 193–205. [CrossRef]
27. Pavlič, P.; Žerovnik, J. Roman domination number of the Cartesian products of paths and cycles. *Electron. J. Comb.* **2012**, *16*, P19. [CrossRef]
28. Steimle, A.; Staton, W. The isomorphism classes of the generalized Petersen graphs. *Discret. Math.* **2009**, *309*, 231–237. [CrossRef]
29. Watkins, M.E. A theorem on Tait colorings with an application to the generalized Petersen graphs. *J. Comb. Theory* **1969**, *6*, 152–164. [CrossRef]
30. Behzad, A.; Behzad, M.; Praeger, C.E. On the domination number of the generalized Petersen graphs. *Discret. Math.* **2008**, *308*, 603–610. [CrossRef]
31. Ebrahimi, B.J.; Jahanbakht, N.; Mahmoodian, E.S. Vertex domination of generalized Petersen graphs. *Discret. Math.* **2009**, *309*, 4355–4361. [CrossRef]
32. Fu, X.; Yang, Y.; Jiang, B. On the domination number of generalized Petersen graphs $P(n,2)$. *Discret. Math.* **2009**, *309*, 2445–2451. [CrossRef]
33. Liu, J.; Zhang, X. The exact domination number of generalized Petersen graphs $P(n,k)$ with $n = 2k$ and $n = 2k + 2$. *Comput. Appl. Math.* **2014**, *33*, 497–506. [CrossRef]
34. Tong, C.; Lin, X.; Yang, Y.; Luo, M. 2-rainbow domination of generalized Petersen graphs $P(n,2)$. *Discret. Appl. Math.* **2009**, *157*, 1932–1937. [CrossRef]
35. Xu, G. 2-rainbow domination in generalized Petersen graphs $P(n,3)$. *Discret. Appl. Math.* **2009**, *157*, 2570–2573. [CrossRef]
36. Yan, H.; Kang, L.; Xu, G. The exact domination number of the generalized Petersen graphs. *Discret. Math.* **2009**, *309*, 2596–2607. [CrossRef]
37. Zhao, W.; Zheng, M.; Wu, L. Domination in the generalized Petersen graph $P(ck,k)$. *Util. Math.* **2010**, *81*, 157–163.
38. Wang, H.; Xu, X.; Yang, Y. On the Domination Number of Generalized Petersen Graphs $P(ck,k)$. *ARS Comb.* **2015**, *118*, 33–49.
39. Gross, J.L.; Tucker, T.W. *Topological Graph Theory*; Wiley-Interscience: New York, NY, USA, 1987.
40. Malnič, A.; Pisanski, T.; Žitnik, A. The clone cover. *Ars Math. Contemp.* **2015**, *8*, 95–113. [CrossRef]
41. Gabrovšek, B.; Peperko, A.; Žerovnik, J. Independent Rainbow Domination Numbers of Generalized Petersen Graphs P(n,2) and $P(n,3)$. *Mathematics* **2020**, *8*, 996. [CrossRef]

Article

More on Sombor Index of Graphs

Wenjie Ning [1], Yuheng Song [1] and Kun Wang [2,*]

[1] College of Science, China University of Petroleum (East China), Qingdao 266580, China; 20180007@upc.edu.cn (W.N.); s20090028@s.upc.edu.cn (Y.S.)
[2] College of Mathematics and Systems Science, Shandong University of Science and Technology, Qingdao 266590, China
* Correspondence: skd996195@sdust.edu.cn

Abstract: Recently, a novel degree-based molecular structure descriptor, called Sombor index was introduced. Let $G = (V(G), E(G))$ be a graph. Then, the Sombor index of G is defined as $SO(G) = \sum_{uv \in E(G)} \sqrt{d_G^2(u) + d_G^2(v)}$. In this paper, we give some lemmas that can be used to compare the Sombor indices between two graphs. With these lemmas, we determine the graph with maximum SO among all cacti with n vertices and k cut edges. Furthermore, the unique graph with maximum SO among all cacti with n vertices and p pendant vertices is characterized. In addition, we find the extremal graphs with respect to SO among all quasi-unicyclic graphs.

Keywords: topological index; vertex degree; Sombor index; cactus; quasi-unicyclic graph

1. Introduction

In this paper, we only consider simple undirected graphs. Let $G = (V(G), E(G))$ be a graph with n vertices and m edges. If $m = n + k - 1$, then G is called a *k-cyclic graph*. A 1-cyclic graph is usually called a *unicyclic graph*. The *complement* G^c of G is the graph with the vertex set $V(G)$, and $xy \in E(G^c) \Leftrightarrow xy \notin E(G)$. The *degree* of a vertex v in G, denoted by $d_G(v)$, is the number of edges incident with v. A vertex of degree one is called a *pendant vertex* of G, while the edge incident with a pendant vertex is known as a *pendant edge*. The vertex adjacent to a pendant vertex is usually called a *support vertex*. To *subdivide* an edge e is to delete e, add a new vertex x, and join x to the end-vertices of e. Suppose $D \subseteq E(G)$. Then, denote by $G - D$ the graph obtained from G by deleting all the elements in D. If $D = \{e\}$, we write $G - e$ for $G - \{e\}$ for simplicity. For a connected graph G, if $G - e$ is disconnected, then e is called a *cut edge*. If $D \subseteq E(G^c)$, denote by $G + D$ the graph obtained from G by adding all of elements in D to the graph G.

A graph invariant is a numerical quantity which is invariant under graph isomorphism. It is usually referred to as a topological index in chemical graph theory. It is shown that some topological indices can be used to reflect physico-chemical and biological properties of molecules in quantitative structure–activity relationship (QSAR) and quantitative structure-property relationship (QSPR) studies [1–3]. Among various topological indices, degree-based and distance-based topological indices have been extensively investigated (see in [4–7]).

In 2021, a novel degree-based topological index was introduced by I. Gutman in [8], called the *Sombor index*. It was inspired by the geometric interpretation of degree-radii of the edges and defined as

$$SO(G) = \sum_{uv \in E(G)} \sqrt{d_G^2(u) + d_G^2(v)}$$

for a graph G. I. Gutman [8] also defined the *reduced Sombor index* as

$$SO_{red}(G) = \sum_{uv \in E(G)} \sqrt{(d_G(u) - 1)^2 + (d_G(v) - 1)^2}.$$

Later, K. C. Das et al. [9] proposed the following index:

$$SO^{\ddagger}(G) = \sum_{uv \in E(G)} \sqrt{(d_G(u)+1)^2 + (d_G(v)+1)^2}.$$

We name it as the *increased Sombor index* in this paper.

Recently, the Sombor index has received a lot of attention within mathematics and chemistry. For example, the chemical applicability of the Sombor index, especially the predictive and discriminative potentials was investigated in [10,11]. The results indicate that the Sombor index may be successfully applied for the modeling of thermodynamic properties of compounds and confirm the suitability of this new index in QSPR analysis. For more chemical applications, the readers may see in [12–14] for reference. K. C. Das et al. [15,16] obtained some lower and upper bounds on SO in terms of graph parameters. They also presented some relations between SO and the Zagreb indices. The relations between SO and other degree-based indices were examined in [17]. Graphs having maximum Sombor index among all connected k-cyclic graphs of order n, where $1 \leq k \leq n-2$, were investigated in [9,18]. R. Cruz et al. [19] characterized the extremal graphs with respect to SO over all (connected) chemical graphs, chemical trees, and hexagonal systems. H. Liu [20] determined the extremal graphs with maximum SO among all cacti with fixed number of cycles and perfect matchings. N. Ghanbari et al. [21] studied this index for certain graphs and also examined the effects on $SO(G)$ when G is modified by operations on vertex and edge of G. Inspired by these works, we establish some new extremal results of the Sombor index.

Recall that a connected graph is a *cactus* if any two of its cycles have at most one common vertex. A connected graph G is called a *quasi-unicyclic graph* if there is a vertex $v \in V(G)$ such that $G - v$ is unicyclic. Let G_1 and G_2 be two graphs with no vertices in common. The *join* of G_1 and G_2, denoted by $G_1 \vee G_2$, is the graph with $V(G_1 \vee G_2) = V(G_1) \cup V(G_2)$ and $E(G_1 \vee G_2) = E(G_1) \cup E(G_2) \cup \{x_1 x_2 : x_1 \in V(G_1), x_2 \in V(G_2)\}$. Let P_n, C_n and S_n be the path, cycle and the star with n vertices, respectively.

This paper is organized as follows. In Section 2, some lemmas are introduced to compare the Sombor indices between two graphs. As applications, in Section 3, the unique graph with maximum SO among all cacti with n vertices and k cut edges is determined. Furthermore, the unique graph with maximum SO among all cacti with n vertices and p pendant vertices is characterized. In Section 4, we present the minimum and maximum SO of quasi-unicyclic graphs.

2. Preliminaries

For convenience, let $f(x, y) = \sqrt{x^2 + y^2}$, where $x, y \geq 1$. For $f(x, y)$, we have the following result.

Lemma 1 ([20,22]). *Let $h(x, y)$ be defined for $x \geq 1, y \geq 1$ as*

$$h(x, y) = f(x+1, y) - f(x, y) = \sqrt{(x+1)^2 + y^2} - \sqrt{x^2 + y^2}.$$

Then, for any value of $y \geq 1$, h is increasing as a function of x; for any value of $x \geq 1$, h is decreasing as a function of y.

Let $P = uu_1 \cdots u_k$ be a path in a graph G with $d_G(u) \geq 3$, $d_G(u_1) = \cdots = d_G(u_{k-1}) = 2$ and $d_G(u_k) = 1$. Then, P is called a *pendant path* in G and u is called the *origin* of P. In [23], B. Horoldagva et al. showed the following transformation.

Lemma 2 ([23]). *Let P and Q be two pendant paths with origins u and v in graph G, respectively. Let x be a neighbor vertex of u who lies on P and y be the pendant vertex on Q. Denote $G' = (G - ux) + xy$. Then, $SO(G) > SO(G')$.*

Now, we introduce some new transformations which increase the Sombor index of a graph.

Lemma 3. *Let G be a graph and $e = uv$ an edge of G with $N_G(u) \cap N_G(v) = \emptyset$. Let G' be the graph obtained from G by first deleting the edge e and identifying u with v, and then attaching a pendant vertex w to the common vertex (see Figure 1). If $d_G(u) \geq 2$ and $d_G(v) \geq 2$, then $SO(G) < SO(G')$.*

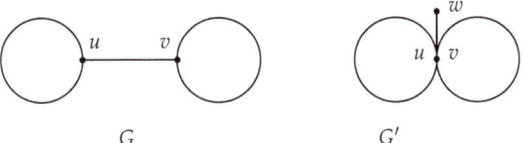

Figure 1. Graphs G and G'.

Proof of Lemma 3. Suppose $d_G(u) = p+1$, $d_G(v) = q+1$, $N_G(u) = \{v, u_1, u_2, \ldots, u_p\}$ and $N_G(v) = \{u, v_1, v_2, \ldots, v_q\}$. Then, $p, q \geq 1$. Therefore,

$$\begin{aligned}
SO(G') - SO(G) &= f(p+q+1,1) - f(p+1,q+1) \\
&+ \sum_{i=1}^{p}[f(p+q+1, d_{G'}(u_i)) - f(p+1, d_G(u_i))] \\
&+ \sum_{i=1}^{q}[f(p+q+1, d_{G'}(v_i)) - f(q+1, d_G(v_i))] \\
&> f(p+q+1,1) - f(p+1,q+1) \\
&> 0.
\end{aligned}$$

□

Lemma 4. *Let G be a graph and $G_{u,v}(p,q)$ the graph obtained from G by attaching p and q pendant edges to u and v, respectively, where $u, v \in V(G)$ and $p \geq q \geq 1$. Suppose $|N_G(u) \setminus \{v\}| = |N_G(v) \setminus \{u\}| = a$. Let $N_G(u) \setminus \{v\} = \{x_1, x_2, \ldots, x_a\}$ and $N_G(v) \setminus \{u\} = \{y_1, y_2, \ldots, y_a\}$. If $d_G(x_i) = d_G(y_i)$ for each $1 \leq i \leq a$, then $SO(G_{u,v}(p,q)) < SO(G_{u,v}(p+1, q-1))$.*

Proof of Lemma 4. If $uv \in E(G)$, then by Lemma 1,

$$\begin{aligned}
&SO(G_{u,v}(p+1,q-1)) - SO(G_{u,v}(p,q)) \\
={}& f(a+p+2, a+q) - f(a+p+1, a+q+1) \\
&+ \sum_{i=1}^{a}[f(a+p+2, d_G(x_i)) - f(a+p+1, d_G(x_i))] \\
&+ \sum_{i=1}^{a}[f(a+q, d_G(y_i)) - f(a+q+1, d_G(y_i))] \\
&+ (p+1)f(a+p+2,1) - pf(a+p+1,1) + (q-1)f(a+q,1) - qf(a+q+1,1) \\
>{}& \sum_{i=1}^{a}[h(a+p+1, d_G(x_i)) - h(a+q, d_G(y_i))] + f(a+p+2,1) - f(a+q,1) \\
&+ p[f(a+p+2,1) - f(a+p+1,1)] - q[f(a+q+1,1) - f(a+q,1)] \\
\geq{}& f(a+p+2,1) - f(a+q,1) + ph(a+p+1,1) - qh(a+q,1) \\
>{}& 0.
\end{aligned}$$

Now, suppose $uv \notin E(G)$. Then $d_G(u) = d_G(v) = a$. Therefore,

$$
\begin{aligned}
&SO(G_{u,v}(p+1,q-1)) - SO(G_{u,v}(p,q)) \\
&= \sum_{i=1}^{a}[f(a+p+1,d_G(x_i)) - f(a+p,d_G(x_i))] \\
&\quad + \sum_{i=1}^{a}[f(a+q-1,d_G(y_i)) - f(a+q,d_G(y_i))] \\
&\quad +(p+1)f(a+p+1,1) - pf(a+p,1) + (q-1)f(a+q-1,1) - qf(a+q,1) \\
&= \sum_{i=1}^{a}[h(a+p,d_G(x_i)) - h(a+q-1,d_G(y_i))] + f(a+p+1,1) - f(a+q-1,1) \\
&\quad + p[f(a+p+1,1) - f(a+p,1)] - q[f(a+q,1) - f(a+q-1,1)] \\
&\geq f(a+p+1,1) - f(a+q-1,1) + ph(a+p,1) - qh(a+q-1,1) \\
&> 0.
\end{aligned}
$$

□

Lemma 5. *Let G be a graph and $C_4 = v_1v_2v_3v_4v_1$ a 4-cycle in G. Suppose $N_G(v_1) \cap N_G(v_3) = \{v_2, v_4\}$. Let $N_G(v_1) \setminus \{v_2, v_4\} = \{v_{11}, v_{12}, \ldots, v_{1n_1}\}$ and $N_G(v_3) \setminus \{v_2, v_4\} = \{v_{31}, v_{32}, \ldots, v_{3n_3}\}$, where $n_1 = d_G(v_1) - 2 > 0$ and $n_3 = d_G(v_3) - 2 > 0$. Let $G' = (G - \{v_1v_{11}, \ldots, v_1v_{1n_1}\}) + \{v_3v_{11}, \ldots, v_3v_{1n_1}\}$. Then, $SO(G) < SO(G')$.*

Proof of Lemma 5. Suppose $d_G(v_2) = n_2 + 2$ and $d_G(v_4) = n_4 + 2$, where $n_2, n_4 \geq 0$. By Lemma 1,

$$
\begin{aligned}
&SO(G') - SO(G) \\
&= \sum_{i=1}^{n_1}[f(n_1+n_3+2,d_G(v_{1i})) - f(n_1+2,d_G(v_{1i}))] \\
&\quad + \sum_{j=1}^{n_3}[f(n_1+n_3+2,d_G(v_{3j})) - f(n_3+2,d_G(v_{3j}))] \\
&\quad + f(n_1+n_3+2,n_2+2) - f(n_3+2,n_2+2) - [f(n_1+2,n_2+2) - f(2,n_2+2)] \\
&\quad + f(n_1+n_3+2,n_4+2) - f(n_3+2,n_4+2) - [f(n_1+2,n_4+2) - f(2,n_4+2)] \\
&> f(n_1+n_3+2,n_2+2) - f(n_3+2,n_2+2) - [f(n_1+2,n_2+2) - f(2,n_2+2)] \\
&\quad + f(n_1+n_3+2,n_4+2) - f(n_3+2,n_4+2) - [f(n_1+2,n_4+2) - f(2,n_4+2)] \\
&= \sum_{i=2}^{n_1+1}[h(n_3+i,n_2+2) - h(i,n_2+2)] + \sum_{i=2}^{n_1+1}[h(n_3+i,n_4+2) - h(i,n_4+2)] \\
&> 0.
\end{aligned}
$$

□

Lemma 6. *Let G be a graph and $C_3 = v_1v_2v_3v_1$ be a 3-cycle in G. Suppose $N_G(v_1) \cap N_G(v_2) = \{v_3\}$. Let $N_G(v_1) \setminus \{v_2, v_3\} = \{v_{11}, v_{12}, \ldots, v_{1n_1}\}$ and $N_G(v_2) \setminus \{v_1, v_3\} = \{v_{21}, v_{22}, \ldots, v_{2n_2}\}$, where $n_1 = d_G(v_1) - 2 > 0$ and $n_2 = d_G(v_2) - 2 > 0$. Denote $G' = (G - \{v_1v_{11}, \ldots, v_1v_{1n_1}\}) + \{v_2v_{11}, \ldots, v_2v_{1n_1}\}$. Then, $SO(G) < SO(G')$.*

Proof of Lemma 6. Suppose $d_G(v_3) = n_3 + 2$, where $n_3 \geq 0$. By Lemma 1,

$$
\begin{aligned}
SO(G') - SO(G) &= \sum_{i=1}^{n_1}[f(n_1+n_2+2,d_G(v_{1i})) - f(n_1+2,d_G(v_{1i}))] \\
&+ \sum_{j=1}^{n_2}[f(n_1+n_2+2,d_G(v_{2j})) - f(n_2+2,d_G(v_{2j}))] \\
&+ f(n_1+n_2+2,2) - f(n_1+2,n_2+2) \\
&+ f(n_1+n_2+2,n_3+2) - f(n_2+2,n_3+2) \\
&- [f(n_1+2,n_3+2) - f(2,n_3+2)] \\
&> f(n_1+n_2+2,n_3+2) - f(n_2+2,n_3+2) \\
&- [f(n_1+2,n_3+2) - f(2,n_3+2)] \\
&= \sum_{i=2}^{n_1+1}[h(n_2+i,n_3+2) - h(i,n_3+2)] \\
&> 0.
\end{aligned}
$$

□

Remark 1. *By the definitions of SO and $SO^\ddagger$, it is easy to see that Lemmas 3–6 also hold if we replace SO with $SO^\ddagger$ in these lemmas.*

By Remark 1, we easily get the following result which will be used later.

Theorem 1. *Let G be the graph with maximum increased Sombor index among all unicyclic graphs of order n. Then $G \cong C_3(n-3,0,0)$, where $C_3(n-3,0,0)$ is obtained from a 3-cycle C_3 by attaching $n-3$ pendant vertices to one vertex of C_3.*

Proof of Theorem 1. By considering the version of $SO^\ddagger$ of Lemma 3, each cut edge of G is pendant and the girth of G is 3. Moreover, all pendant vertices are adjacent to one common vertex by Lemma 6. Therefore, $G \cong C_3(n-3,0,0)$. □

3. Sombor Index of Cacti

Denote by C_n^k the set of all cacti of order n with k cut edges, and $C(n,p)$ the set of all cacti of order n with p pendant vertices. Then, $0 \le k \le n-1$ and $k \ne n-2$. In this section, we investigate the maximal values of the Sombor index over the sets C_n^k and $C(n,p)$.

Given two integers k and n with $0 \le k \le n-1$ and $k \ne n-2$, if $n-k$ is even, denote by G_1 the graph obtained from a star S_{n-1} by first adding $\frac{n-k-2}{2}$ new edges between its pendant vertices such that no two of the new edges are adjacent, and then subdividing one new edge; if $n-k$ is odd, denote by G_2 the graph obtained from a star S_n by adding $\frac{n-k-1}{2}$ new edges between its pendant vertices such that no two of the new edges are adjacent (see Figure 2).

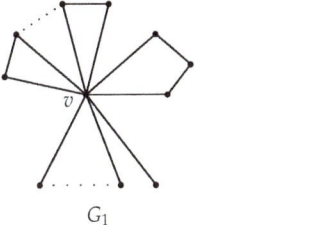

Figure 2. Graphs G_1 and G_2.

The following theorem shows that the graph S_n has maximum Sombor index over C_n^{n-1}. Therefore, we assume $k \ne n-1$ in the following.

Theorem 2 ([8]). *For any tree T of order n,*

$$SO(P_n) \leq SO(T) \leq SO(S_n).$$

Equality holds if and only if $T \cong P_n$ *or* $T \cong S_n$.

Theorem 3. *For graphs in* $\mathcal{C}_n^k$, *where* $0 \leq k \leq n-3$,
(1) *if* $n-k$ *is even,* G_1 *is the unique graph with maximum Sombor index;*
(2) *if* $n-k$ *is odd,* G_2 *is the unique graph with maximum Sombor index, where* G_1 *and* G_2 *are depicted in Figure* 2.

Proof of Theorem 3. Let G be the graph with maximum Sombor index in $\mathcal{C}_n^k$. By Lemma 3, each cut edge of G is pendant. We show that the following propositions hold for G. □

Proposition 1. *Each cycle in G is of length 3 or 4.*

Proof of Proposition 1. Suppose to the contrary that G has a cycle $C = v_1 v_2 \cdots v_t v_1$ of length $t \geq 5$. Without loss of generality, we assume that $d_G(v_1) = \max\{d_G(v_i) | 1 \leq i \leq t\}$. Define $G' = (G - v_3 v_4) + \{v_1 v_3, v_1 v_4\}$. Then, $G' \in \mathcal{C}_n^k$ and

$$
\begin{aligned}
SO(G') &- SO(G) \\
&> f(d_{G'}(v_1), d_{G'}(v_3)) + f(d_{G'}(v_1), d_{G'}(v_4)) - f(d_G(v_3), d_G(v_4)) \\
&= f(d_G(v_1) + 2, d_G(v_3)) + f(d_G(v_1) + 2, d_G(v_4)) - f(d_G(v_3), d_G(v_4)) \\
&> 0,
\end{aligned}
$$

a contradiction to the choice of G. Therefore, Proposition 1 holds. □

Proposition 2. *Each 4-cycle has at most one vertex of degree larger than 2 in G.*

Proof of Proposition 2. Suppose there is a 4-cycle $C = v_1 v_2 v_3 v_4 v_1$ containing at least two vertices of degree larger than 2. Since G has maximum Sombor index, any two vertices of C with degree larger than 2 must be adjacent by Lemma 5. Thus, C contains exactly two adjacent vertices of degree 2. Without loss of generality, we assume that $d_G(v_1) = n_1 + 2 > 2, d_G(v_2) = n_2 + 2 > 2$ and $d_G(v_3) = d_G(v_4) = 2$. Suppose $N_G(v_1) \setminus \{v_2, v_4\} = \{v_{11}, v_{12}, \ldots, v_{1n_1}\}$ and $N_G(v_2) \setminus \{v_1, v_3\} = \{v_{21}, v_{22}, \ldots, v_{2n_2}\}$. Let $G' = (G - \{v_2 v_{21}, v_2 v_{22}, \ldots, v_2 v_{2n_2}\}) + \{v_1 v_{21}, v_1 v_{22}, \ldots, v_1 v_{2n_2}\}$. Then, $G' \in \mathcal{C}_n^k$ and

$$
\begin{aligned}
SO(G') - SO(G) &> f(n_1 + n_2 + 2, 2) - f(n_1 + 2, n_2 + 2) \\
&\quad + f(n_1 + n_2 + 2, 2) - f(n_1 + 2, 2) - [f(n_2 + 2, 2) - f(2, 2)] \\
&= f(n_1 + n_2 + 2, 2) - f(n_1 + 2, n_2 + 2) + \sum_{i=2}^{n_2+1} [h(n_1 + i, 2) - h(i, 2)] \\
&> 0,
\end{aligned}
$$

a contradiction.

By Lemma 6, each 3-cycle has at most one vertex of degree larger than 2. Combining it with Propositions 1 and 2, G is obtained from s copies of C_4 and t copies of C_3 by first taking one vertex of each of them and fusing them together into a new common vertex v, and then attaching k pendant vertices at v, where $3s + 2t + k + 1 = n$. Suppose $s \geq 2$. Then, there are at least two 4-cycles $C = v x_1 y_1 z_1 v$ and $C' = v x_2 y_2 z_2 v$. Let $G' = (G - \{x_1 y_1, x_2 y_2\}) + \{x_1 x_2, y_1 v, y_2 v\}$. Then $G' \in \mathcal{C}_n^k$ and $SO(G') - SO(G) > (6f(d_G(v) + 2, 2) + 3f(2, 2)) - (4f(d_G(v), 2) + 4f(2, 2)) > 0$, a contradiction. This implies $0 \leq s \leq 1$. Therefore, $s = 1$ if $n - k$ is even and $s = 0$ otherwise, i.e., $G \cong G_1$ if $n - k$ is even and $G \cong G_2$ otherwise. □

Next, we find the maximal graph with respect to the Sombor index among $\mathcal{C}(n, p)$ with n vertices and p pendant vertices. As S_{n-1} is the only graph with $n-1$ pendant vertices, we assume $0 \leq p \leq n-2$ in the following. Before we give our main result, we show a lemma.

Lemma 7. *Let G be a graph and $e = uv \in E(G)$ with $N_G(u) \cap N_G(v) = \emptyset$. Suppose $N_G(u) \setminus \{v\} = \{x_1, \ldots, x_s, w_1, \ldots, w_p\}$ and $N_G(v) \setminus \{u\} = \{y_1, \ldots, y_t, z_1, \ldots, z_q\}$, where $d(x_1) = \cdots = d(x_s) = 2$, $d(w_1) = \cdots = d(w_p) = 1$, $d(y_1) = \cdots = d(y_t) = 2$ and $d(z_1) = \cdots = d(z_q) = 1$. Suppose $s \geq 2$. Let G' be obtained from G by first deleting the edge e and identifying u with v, and then subdividing the edge ux_1. If $t + q \geq 2$, or $t = 0$ with $q = 1$, then $SO(G) < SO(G')$.*

Proof of Lemma 7. By direct calculation, we get

$$
\begin{aligned}
& SO(G') - SO(G) \\
= \ & (p+q)f(s+p+t+q,1) - pf(s+p+1,1) - qf(t+q+1,1) \\
& + (s+t)f(s+p+t+q,2) - sf(s+p+1,2) - tf(t+q+1,2) \\
& + f(2,2) - f(t+q+1,s+p+1) \\
\geq \ & q[f(s+p+t+q,1) - f(t+q+1,1)] + s[f(s+p+t+q,2) - f(s+p+1,2)] \\
& + t[f(s+p+t+q,2) - f(t+q+1,2)] \\
& - [f(t+q+1,s+p+1) - f(2,s+p+1) + f(s+p+1,2) - f(2,2)] \quad (1) \\
= \ & q \sum_{i=2}^{s+p} h(t+q+i-1,1) + t\sum_{i=2}^{s+p} h(t+q+i-1,2) - \sum_{i=2}^{s+p} h(i,2) \\
& + s[f(s+p+t+q,2) - f(s+p+1,2)] \\
& - [f(t+q+1,s+p+1) - f(2,s+p+1)].
\end{aligned}
$$

We denote the right side of equation (1) by A. Then, by Lemma 1,

$$
\begin{aligned}
A \ & > \ s[f(s+p+t+q,2) - f(s+p+1,2)] \\
& \quad - [f(t+q+1,s+p+1) - f(2,s+p+1)] \\
& = \ s\sum_{i=2}^{t+q} h(s+p+i-1,2) - \sum_{i=2}^{t+q} h(i,s+p+1) \\
& > \ 0
\end{aligned}
$$

if $t + q \geq 2$. Now, suppose $t = 0$ and $q = 1$. Then, $A = \sum_{i=2}^{s+p}[h(i,1) - h(i,2)] > 0$ by Lemma 1, which completes the proof. □

Denote by $DS_{p,q}$ the double star obtained from a star S_{p+2} by attaching q pendant vertices to one pendant vertex. Then $DS_{p,q}$ has $n = p + q + 2$ vertices.

Theorem 4. *For graphs in $\mathcal{C}(n,p)$, where $0 \leq p \leq n-2$ and $n \geq 5$,*
(1) if $p = n-2$, $DS_{n-3,1}$ is the unique graph with maximum Sombor index;
(2) if $p \leq n-3$ and $n - p$ is even, G_1 is the unique graph with maximum Sombor index;
(3) if $p \leq n-3$ and $n - k$ is odd, G_2 is the unique graph with maximum Sombor index, where G_1 and G_2 are depicted in Figure 2.

Proof of Theorem 4. If $p = n-2$, then any graph in $\mathcal{C}(n,p)$ is a double star. By Lemma 4, $DS_{n-3,1}$ is the unique graph with maximum Sombor index. Now suppose $p \leq n-3$. Let G be the graph with maximum Sombor index in $\mathcal{C}(n,p)$. Then, the following claims hold. □

Claim 1. *G must contain a cycle.*

Proof of Claim 1. Suppose not. Then, G is a tree. As $p \leq n-3$, there are two non-pendant vertices u and v of G with $uv \notin E(G)$. Let $G' = G + uv$. Then, $G' \in \mathcal{C}(n, p)$ and $SO(G') > SO(G)$, a contradiction to the choice of G.

By the same argument as that of Theorem 3, each cycle in G is of length 3 or 4. Moreover, each cycle of G has at most one vertex of degree larger than 2. Let T be the graph obtained from G by deleting all vertices of degree 2 in each cycle. Then, T is a tree. Let $d(T)$ be the diameter of T. □

Claim 2. $d(T) \leq 3$.

Proof of Claim 2. Suppose $d(T) \geq 4$. Then there are two non-pendant vertices u and v of T with $uv \notin E(T)$. Let $G' = G + uv$. Then $G' \in \mathcal{C}(n, p)$ and $SO(G') > SO(G)$, a contradiction, which implies Claim 2 holds.

Similarly, we have □

Claim 3. *For any two cycles C_1 and C_2 in G, the length of the path connecting them is 0 or 1.*

Proof of Claim 3. Suppose there are two cycles C_1 and C_2 such that the length of the path P connecting them is larger than 1. Let $P = u_0 u_1 \cdots u_l$, where $u_0 \in V(C_1)$ and $u_l \in V(C_2)$. Then $l > 1$. Define $G' = G + u_0 u_l$. Then $G' \in \mathcal{C}(n, p)$ and $SO(G') > SO(G)$, a contradiction, which implies Claim 3 holds. □

Claim 4. *Any two cycles of G have one common vertex.*

Proof of Claim 4. Suppose there are two cycles C_1 and C_2 having no vertices in common. Then by Claim 3, the length of the path P connecting them is 1. Suppose $P = u_1 v_1$, where $u_1 \in V(C_1)$ and $v_1 \in V(C_2)$. As each cycle of G has at most one vertex of degree larger than 2, $N_G(u_1) \cap N_G(v_1) = \emptyset$.

We show that for each vertex $w \in (N_G(u_1) \setminus \{v_1\}) \cup (N_G(v_1) \setminus \{u_1\})$, $d_G(w) \in \{1, 2\}$. Without loss of generality, suppose there is a vertex $w \in N_G(v_1) \setminus \{u_1\}$ with $d_G(w) \geq 3$. Then $w \in V(T)$. Let $G' = G + u_1 w$. Then, $G' \in \mathcal{C}(n, p)$ and $SO(G') > SO(G)$, a contradiction.

Now let $u_2 \in V(C_1) \cap N_G(u_1)$. Let G' be obtained from G by first deleting the edge $u_1 v_1$ and identifying u_1 with v_1, and then subdividing the edge $u_1 u_2$. Then $G' \in \mathcal{C}(n, p)$ and $SO(G') > SO(G)$ by Lemma 7, a contradiction. Therefore, Claim 4 holds. □

Claim 5. *All pendant vertices in G are adjacent to u, where u is the common vertex of all cycles in G.*

Proof of Claim 5. Suppose there is a support vertex $v \neq u$. Let $d_G(u, v)$ be the distance between u and v in G. Then $d_G(u, v) \in \{1, 2\}$ by Claim 2. If $d_G(u, v) = 2$, by letting $G' = G + uv$, we get $G' \in \mathcal{C}(n, p)$ and $SO(G') > SO(G)$, a contradiction. Therefore, $d_G(u, v) = 1$.

Suppose for each vertex $w \in N_G(v) \setminus \{u\}$, $d_G(w) = 1$. Then by Claim 1, there are at least two vertices of degree 2 adjacent to u except v. Let $u_1 \in N_G(u) \setminus \{v\}$ with $d_G(u_1) = 2$. Denote by G' the graph obtained from G by first deleting the edge uv and identifying u with v, and then subdividing the edge uu_1. Then $G' \in \mathcal{C}(n, p)$ and $SO(G') > SO(G)$ by Lemma 7, a contradiction. Therefore, we may assume that there is a vertex $v_1 \in N_G(v) \setminus \{u\}$ with $d_G(v_1) \geq 2$. Let $G' = G + uv_1$. Then $G' \in \mathcal{C}(n, p)$ and $SO(G') > SO(G)$, a contradiction. This completes the proof of Claim 5.

By the same argument as that of Theorem 3, there is at most one 4-cycle in G. Therefore, G is obtained from s copies of C_4 and t copies of C_3 by first taking one vertex of each of them and fusing them together into a new common vertex u, and then attaching p pendant vertices at u, where $3s + 2t + p + 1 = n$ and $0 \leq s \leq 1$. Therefore, $s = 1$ if $n - p$ is even and $s = 0$ otherwise, i.e., $G \cong G_1$ if $n - p$ is even and $G \cong G_2$ otherwise. □

4. Sombor Index of Quasi-Unicyclic Graphs

Let $\mathcal{QU}(n)$ be the set of all quasi-unicyclic graphs of order n. Denote by $\infty(p,l,q)$ the graph obtained from two cycles C_p and C_q by connecting a vertex $u \in V(C_p)$ and a vertex $v \in V(C_q)$ by a path $v_0 v_1 \cdots v_l$ of length l (identifying u with v if $l = 0$), where $v_0 = u$, $v_l = v$ and $p + q + l = n + 1$. Let $\Theta(s,t,r)$ be a union of three paths $P_{s+1}, P_{t+1}, P_{r+1}$ resp. with common end vertices, where $s + t + r + 1 = n$, $s \geq t \geq r \geq 1$ and at most one of them is 1.

Theorem 5. *Let $G \in \mathcal{QU}(n)$ be the graph with minimum Sombor index, where $n \geq 4$. Then $G \cong \Theta(s,t,1)$ or $\infty(p,1,q)$, and $SO(G) = (2n-5)\sqrt{2} + 4\sqrt{13}$.*

Proof of Theorem 5. As $G \in \mathcal{QU}(n)$, there is a vertex v_n in G such that $G - v_n = H$ is a unicyclic graph. Let $d_G(v_n) = k$. Then $k \geq 2$. As G has the minimum Sombor index, $k = 2$. Let $N_G(v_n) = \{v_1, v_2\}$.

We first show that H has at most two pendant vertices. Suppose H has at least three pendant vertices. If $\max\{d_H(v_1), d_H(v_2)\} \geq 2$, then G has at least two pendant paths, say $P = uu_1 \cdots u_s$ and $Q = ww_1 \cdots w_t$, where $d_G(u), d_G(w) \geq 3$. Let $G' = (G - uu_1) + u_1 w_t$. Then, $G' \in \mathcal{QU}(n)$ and $SO(G') < SO(G)$ by Lemma 2, a contradiction. Therefore, we may assume that $d_H(v_1) = d_H(v_2) = 1$. Then, there is a pendant path $P = xx_1 \cdots x_l$ in G, where $d_G(x) = a \geq 3$, $d_G(x_1) = \cdots = d_G(x_{l-1}) = 2$ and $d_G(x_l) = 1$. Define $G' = (G - \{xx_1, v_1 v_n\}) + \{v_1 x_1, x_l v_n\}$. Then, $G' \in \mathcal{QU}(n)$. By direct calculation, we get $SO(G') - SO(G) < f(2,2) - f(1,a) < 0$ if $l = 1$, and

$$\begin{aligned} SO(G') - SO(G) &< 2f(2,2) - (f(a,2) + f(1,2)) \\ &= f(2,2) - f(1,2) - [f(a,2) - f(2,2)] \\ &= h(1,2) - \sum_{i=2}^{a-1} h(i,2) \\ &< 0 \end{aligned}$$

if $l \geq 2$. This contradicts to the definition of G. Therefore, there are at most two pendant vertices in H.

Now, we show that every pendant vertex in H is adjacent to v_n in G. Suppose there is a pendant vertex v in H with $v \notin \{v_1, v_2\}$. Then there is one vertex in $\{v_1, v_2\}$, say v_1, such that $d_G(v_1) = a \geq 3$. Let w be the neighbor of v in G. Obviously, $d_G(w) = b \geq 2$. If $w \in \{v_1, v_2\}$, let $G' = (G - wv_n) + vv_n$. Then, $G' \in \mathcal{QU}(n)$ and $SO(G') < SO(G)$ by Lemma 3, a contradiction. Therefore, $w \neq v_1, v_2$. Now, let $G' = (G - v_n v_1) + v_n v$. Then, $G' \in \mathcal{QU}(n)$ and

$$\begin{aligned} SO(G') - SO(G) &< f(2,2) + f(2,b) - [f(2,a) + f(1,b)] \\ &= f(2,b) - f(1,b) - [f(a,2) - f(2,2)] \\ &= h(1,b) - \sum_{i=2}^{a-1} h(i,2) \\ &< 0, \end{aligned}$$

a contradiction.

From the above, G is a bicyclic graph with no pendant vertices. i.e., $G \cong \Theta(s,t,r)$ or $\infty(p,l,q)$. By direct calculation, we have $SO(\Theta(s,t,r)) = (n-5)f(2,2) + 6f(2,3)$ if $r \geq 2$, $SO(\Theta(s,t,r)) = (n-4)f(2,2) + 4f(2,3) + f(3,3)$ if $r = 1$, $SO(\infty(p,l,q)) = (n-3)f(2,2) + 4f(2,4)$ if $l = 0$, $SO(\infty(p,l,q)) = (n-5)f(2,2) + 6f(2,3)$ if $l \geq 2$ and $SO(\infty(p,l,q)) = (n-4)f(2,2) + 4f(2,3) + f(3,3)$ if $l = 1$. Since $(n-3)f(2,2) + 4f(2,4) > (n-5)f(2,2) + 6f(2,3) > (n-4)f(2,2) + 4f(2,3) + f(3,3)$, we get $G \cong \Theta(s,t,1)$ or $\infty(p,1,q)$, and $SO(G) = (n-4)f(2,2) + 4f(2,3) + f(3,3) = (2n-5)\sqrt{2} + 4\sqrt{13}$. □

Here, we recall some theory of majorization.

A subset $X \subseteq \mathbb{R}^n$ is a *convex set* if for any $x, y \in X$ and any λ with $0 < \lambda < 1$, $\lambda x + (1-\lambda)y \in X$. Let $X \subseteq \mathbb{R}^n$ be a convex set. For a function $f: X \to \mathbb{R}$, if for any $x, y \in X$ and any λ with $0 < \lambda < 1$, $f(\lambda x + (1-\lambda)y) \leq \lambda f(x) + (1-\lambda)f(y)$, then f is called a *convex function*. If the inequality above is strict for all $x, y \in X$ with $x \neq y$, then f is a *strictly convex function*. Let I be an interval and $f : I \to \mathbb{R}$ be a real twice-differentiable function on I, then it is well-known that f is convex if and only if $f''(x) \geq 0$ for all $x \in I$, and f is strictly convex if $f''(x) > 0$ for all $x \in I$.

For each vector $x = (x_1, x_2, \ldots, x_n) \in \mathbb{R}^n$, consider the decreasing rearrangement of it, i.e., we always assume that $x_1 \geq x_2 \geq \cdots \geq x_n$. Then we have the following definition and majorization inequality.

Definition 1 ([24]). *For $x, y \in \mathbb{R}^n$,*

$$x \prec y \text{ if } \begin{cases} \sum_{i=1}^{k} x_i \leq \sum_{i=1}^{k} y_i, & k = 1, 2, \ldots, n-1, \\ \sum_{i=1}^{n} x_i = \sum_{i=1}^{n} y_i. \end{cases}$$

When $x \prec y$, x is said to be majorized by y (y majorizes x).

Lemma 8 ([25]). *Let $I \subseteq \mathbb{R}$ be an interval and $f : I \to \mathbb{R}$ a strictly convex function. Let $c = (c_1, c_2, \ldots, c_n)$ and $d = (d_1, d_2, \ldots, d_n)$ be two vectors in $\mathbb{R}^n$ with $c_i, d_i \in I$ for each $i = 1, 2, \ldots, n$. If $c \prec d$, then $\sum_{i=1}^{n} f(c_i) \leq \sum_{i=1}^{n} f(d_i)$, with equality if and only if $c = d$.*

Theorem 6. *Let $G \in \mathcal{QU}(n)$. Then,*

$$SO(G) \leq 2(n-4)\sqrt{(n-1)^2 + 4} + 4\sqrt{(n-1)^2 + 9} + (n+2)\sqrt{2},$$

with equality if and only if $G \cong C_3(n-4, 0, 0) \vee K_1$, where $C_3(n-4, 0, 0)$ is obtained from a 3-cycle C_3 by attaching $n-4$ pendant vertices to one vertex of C_3.

Proof of Theorem 6. Let v_n be a vertex in G such that $G - v_n = H$ is a unicyclic graph. Let $V(H) = \{v_1, v_2, \ldots, v_{n-1}\}$ and $d_G(v_n) = k$. Then, $2 \leq k \leq n-1$. By the definition of the Sombor index, we get

$$SO(G) \leq \sum_{i=1}^{n-1} \sqrt{(n-1)^2 + (d_H(v_i) + 1)^2} + \sum_{v_i v_j \in E(H)} \sqrt{(d_H(v_i) + 1)^2 + (d_H(v_j) + 1)^2}.$$

Moreover, the equality holds if and only if $d_G(v_n) = n-1$. First we consider the maximum value of $\sum_{v_i v_j \in E(H)} \sqrt{(d_H(v_i) + 1)^2 + (d_H(v_j) + 1)^2}$. By Theorem 1,

$$\sum_{v_i v_j \in E(H)} \sqrt{(d_H(v_i) + 1)^2 + (d_H(v_j) + 1)^2} = SO^{\ddagger}(H) \leq SO^{\ddagger}(C_3(n-4, 0, 0)),$$

with equality if and only if $H \cong C_3(n-4, 0, 0)$.

Now, we calculate the maximum value of $\sum_{i=1}^{n-1} \sqrt{(n-1)^2 + (d_H(v_i) + 1)^2}$. Consider the function

$$g(x) = \sqrt{(n-1)^2 + (x+1)^2}, \ 1 \leq x \leq n-2.$$

Then, $g''(x) = \frac{2(n-1)^2+(x+1)^2}{2((n-1)^2+(x+1)^2)^{\frac{3}{2}}} > 0$, $1 \leq x \leq n-2$. Therefore, $g(x)$ is strictly convex on $1 \leq x \leq n-2$. Note that H is a unicyclic graph and $\sum_{i=1}^{n-1} d_H(v_i) = 2(n-1)$, by [26], the degree sequence $d(H) = (d_H(v_1), (d_H(v_2), \ldots, (d_H(v_{n-1})))$ satisfies $d(H) \prec (n-2, 2, 2, \underbrace{1, \ldots, 1}_{n-4})$. By Lemma 8, $\sum_{i=1}^{n-1} g((d_H(v_i)) \leq g(n-2) + 2g(2) + (n-4)g(1)$, that is,

$$\sum_{i=1}^{n-1} \sqrt{(n-1)^2 + ((d_H(v_i)) + 1)^2}$$
$$\leq \sqrt{(n-1)^2 + (n-2+1)^2} + 2\sqrt{(n-1)^2 + (2+1)^2}$$
$$+ (n-4)\sqrt{(n-1)^2 + (1+1)^2}$$
$$= (n-1)\sqrt{2} + 2\sqrt{(n-1)^2 + 9} + (n-4)\sqrt{(n-1)^2 + 4}.$$

Moreover, equality holds if and only if $d(H) = (n-2, 2, 2, \underbrace{1, \ldots, 1}_{n-4})$, i.e., $H \cong C_3(n-4, 0, 0)$.

Based on the above,

$$SO(G)$$
$$\leq (n-1)\sqrt{2} + 2\sqrt{(n-1)^2 + 9} + (n-4)\sqrt{(n-1)^2 + 4}$$
$$+ (n-4)\sqrt{(n-1)^2 + 2^2} + 2\sqrt{(n-1)^2 + 3^2} + \sqrt{3^2 + 3^2}$$
$$= 2(n-4)\sqrt{(n-1)^2 + 4} + 4\sqrt{(n-1)^2 + 9} + (n+2)\sqrt{2}$$

Moreover, the equality holds if and only if $G \cong C_3(n-4, 0, 0) \vee K_1$. □

5. Conclusions

As graph invariants, topological indices are used for QSAR and QSPR studies. Therefore, it is very important to study the extremal graphs with respect to topological indices in chemical graph theory. Until now, many topological indices have been introduced and several of them have been found various applications. As a novel index, the Sombor index has received a lot of attention within mathematics and chemistry. In this paper, we give some transformations to compare the Sombor indices between two graphs. With these transformations, we present the maximum Sombor index among cacti $\mathcal{C}_n^k$ and $\mathcal{C}(n, p)$. Moreover, the maximum and minimum Sombor index among all quasi-unicyclic graphs are characterized. It is interesting to consider the minimum Sombor index of cacti with some graph parameters. We will consider it for future study.

Author Contributions: Methodology, W.N. and Y.S.; investigation, W.N. and Y.S.; writing—original draft preparation, Y.S.; writing—review and editing, W.N. and K.W.; funding acquisition, W.N. All authors have read and agreed to the published version of the manuscript.

Funding: This study was supported by the National Natural Science Foundation of China (Nos. 11801568, 11901150).

Institutional Review Board Statement: Not applicable.

Informed Consent Statement: Not applicable.

Data Availability Statement: Not applicable.

Conflicts of Interest: The authors declare no conflicts of interest.

References

1. Devillers, J.; Balaban, A.T. *Topological Indices and Related Descriptors in QSAR and QSPR*; Gordan and Breach Science Publishers: Amsterdam, The Netherlands, 1999.
2. Karelson, M. *Molecular Descriptors in QSAR/QSPR*; Wiley-Interscience: New York, NY, USA, 2000.
3. Diudea, M.V. *QSPR/QSAR Studies by Molecular Descriptors*; Nova Science Publishers: Huntington, NY, USA, 2000.
4. Gutman, I. Degree-based topological indices. *Croat. Chem. Acta* **2014**, *86*, 351–361. [CrossRef]
5. Xu, K.; Liu, M.; Das, K.C.; Gutman, I.; Furtula, B. A survey on graphs extremal with respect to distance-based topological indices. *MATCH Commun. Math. Comput. Chem.* **2014**, *71*, 461–508.
6. Rada, J.; Cruz, R. Vertex-degree-based topological indices over graphs. *MATCH Commun. Math. Comput. Chem.* **2014**, *72*, 603–616.
7. Das, K.C.; Gutman, I.; Nadjafi-Arani, M.J. Relations between distance-based and degree-based topological indices. *Appl. Math. Comput.* **2015**, *270*, 142–147. [CrossRef]
8. Gutman, I. Geometric approach to degree-based topological indices: Sombor indices. *MATCH Commun. Math. Comput. Chem.* **2021**, *86*, 11–16.
9. Das, K.C.; Ghalavand, A.; Ashrafi, A.R. On a conjecture about the Sombor index of graphs. *arXiv* **2021**, arXiv:2103.17147v1.
10. Redžepović, I. Chemical applicability of Sombor indices. *J. Serb. Chem. Soc.* **2021**, *86*, 445–457. [CrossRef]
11. Deng, H.; Tang, Z.; Wu, R. Molecular trees with extremal values of Sombor indices. *Int. J. Quantum Chem.* **2021**, *121*, e26622. [CrossRef]
12. Aguilar-Sánchez, R.; Méndez-Bermúdez, J.A.; Rodríguez, J.M.; Sigarreta, J.M. Normalized Sombor Indices as Complexity Measures of Random Networks. *Entropy* **2021**, *23*, 976. [CrossRef] [PubMed]
13. Liu, H.; You, L.; Huang, Y. Ordering Chemical Graphs by Sombor Indices and Its Applications. *MATCH Commun. Math. Comput. Chem.* **2022**, *87*, 5–22. [CrossRef]
14. Liu, H.; Chen, H.; Xiao, Q.; Fang, X.; Tang, Z. More on Sombor indices of chemical graphs and their applications to the boiling point of benzenoid hydrocarbons. *Int. J. Quantum Chem.* **2021**, *121*, e26689. [CrossRef]
15. Das, K.C.; Cevik, A.S.; Cangul, I.N.; Shang, Y. On Sombor index. *Symmetry* **2021**, *13*, 140. [CrossRef]
16. Das, K.C.; Shang, Y. Some Extremal Graphs with Respect to Sombor Index. *Mathematics* **2021**, *9*, 1202. [CrossRef]
17. Wang, Z.; Mao, Y.; Li, Y.; Furtula, B. On relations between Sombor and other degree-based indices. *J. Appl. Math. Comput.* **2021**. [CrossRef]
18. Réti, T.; Došlić, T.; Ali, A. On the Sombor index of graphs. *Contrib. Math.* **2021**, *3*, 11–18.
19. Cruz, R.; Gutman, I.; Rada, J. Sombor index of chemical graphs. *Appl. Math. Comput.* **2021**, *399*, 126018. [CrossRef]
20. Liu, H. Maximum Sombor Index Among Cacti. *arXiv* **2021**, arXiv:2103.07924v1.
21. Ghanbari, N.; Alikhani, S. Sombor index of certain graphs. *arXiv* **2021**, arXiv:2102.10409v1.
22. Cruz, R.; Rada, J. Extremal values of the Sombor index in unicyclic and bicyclic graphs. *J. Math. Chem.* **2021**, *59*, 1098–1116. [CrossRef]
23. Horoldagva, B.; Xu, C. On Sombor index of graphs. *MATCH Commun. Math. Comput. Chem.* **2021**, *86*, 701–713.
24. Marshall, A.W.; Olkin, I.; Arnold, B.C. *Inequalities: Theory of Majorization and Its Applications*; Springer: New York, NY, USA, 2011.
25. Karamata, J. Sur une inégalité relative aux fonctions convexes. *Publ. Math. Univ. Belgrade* **1932**, *1*, 145–148.
26. Dimitrov, D.; Ali, A. On the extremal graphs with respect to variable sum exdeg index. *Discrete Math. Lett.* **2019**, *1*, 42–48.

Article

A Proof of a Conjecture on Bipartite Ramsey Numbers $B(2,2,3)$

Yaser Rowshan [1], Mostafa Gholami [1] and Stanford Shateyi [2],*

[1] Department of Mathematics, Institute for Advanced Studies in Basic Sciences (IASBS), Zanjan 66731-45137, Iran; y.rowshan@iasbs.ac.ir (Y.R.); gholami.m@iasbs.ac.ir (M.G.)
[2] Department of Mathematics and Applied Mathematics, School of Mathematical and Natural Sciences, University of Venda, P. Bag X5050, Thohoyandou 0950, South Africa
* Correspondence: stanford.shateyi@univen.ac.za

Abstract: The bipartite Ramsey number $B(n_1, n_2, \ldots, n_t)$ is the least positive integer b, such that any coloring of the edges of $K_{b,b}$ with t colors will result in a monochromatic copy of K_{n_i,n_i} in the i-th color, for some i, $1 \leq i \leq t$. The values $B(2,5) = 17$, $B(2,2,2,2) = 19$ and $B(2,2,2) = 11$ have been computed in several previously published papers. In this paper, we obtain the exact values of the bipartite Ramsey number $B(2,2,3)$. In particular, we prove the conjecture on $B(2,2,3)$ which was proposed in 2015—in fact, we prove that $B(2,2,3) = 17$.

Keywords: Ramsey numbers; bipartite Ramsey numbers; Zarankiewicz number

1. Introduction

The bipartite Ramsey number $B(n_1, n_2, \ldots, n_t)$ is the least positive integer b, such that any coloring of the edges of $K_{b,b}$ with t colors will result in a monochromatic copy of K_{n_i,n_i} in the i-th color, for some i, $1 \leq i \leq t$. The existence of such a positive integer is guaranteed by a result of Erdős and Rado [1].

The Zarankiewicz number $z(K_{m,n}, t)$ is defined as the maximum number of edges in any subgraph G of the complete bipartite graph $K_{m,n}$, such that G does not contain $K_{t,t}$ as a subgraph. Zarankiewicz numbers and related extremal graphs have been studied by many authors, including Kóvari [2], Reiman [3], and Goddard, Henning, and Oellermann in [4].

The study of bipartite Ramsey numbers was initiated by Beineke and Schwenk in 1976 [5], and continued by others, in particular Exoo [6], Hattingh, and Henning [7]. The following exact values have been established: $B(2,5) = 17$ [8], $B(2,2,2,2) = 19$ [9], $B(2,2,2) = 11$ [6]. In the smallest open case for five colors, it is known that $26 \leq B(2,2,2,2,2) \leq 28$ [9]. One can refer to [2,9–14] and it references for further studies. Collins et al. in [8] showed that $17 \leq B(2,2,3) \leq 18$, and in the same source made the following conjecture:

Conjecture 1 ([8]). $B(2,2,3) = 17$.

We intend to get the exact value of the multicolor bipartite Ramsey numbers $B(2,2,3)$. We prove the following result:

Theorem 1. $B(2,2,3) = 17$.

In this paper, we are only concerned with undirected, simple, and finite graphs. We follow [15] for terminology and notations not defined here. Let G be a graph with vertex set $V(G)$ and edge set $E(G)$. The degree of a vertex $v \in V(G)$ is denoted by $\deg_G(v)$, or simply by $\deg(v)$. The neighborhood $N_G(v)$ of a vertex v is the set of all vertices of G adjacent to v and satisfies $|N_G(v)| = \deg_G(v)$. The minimum and maximum degrees of vertices of G are denoted by $\delta(G)$ and $\Delta(G)$, respectively. Additionally, the complete bipartite graph with bipartition (X, Y), where $|X| = m$ and $|Y| = n$, is denoted by $K_{m,n}$. We use $[X, Y]$ to denote the set of edges between the bipartition (X, Y) of G. Let $G = (X, Y)$ be a bipartite graph and $Z \subseteq X$ or $Z \subseteq Y$, the degree sequence of Z denoted by $D_G(Z) = (d_1, d_2, \ldots, d_{|Z|})$, is the list of the degrees of all vertices of Z. The complement of a graph G, denoted by $\overline{G}$, is a graph with same vertices such that two distinct vertices of $\overline{G}$ are adjacent if and only if they are not adjacent in G. H is n-colorable to $(H_1, H_2, \ldots, H_t)$ if there exists a t-coloring of the edges of H such that $H_i \nsubseteq H^i$ for each $1 \leq i \leq t$, where H^i is the spanning subgraph of H with edges of the i-th color.

2. Some Preliminary Results

To prove our main result—namely, Theorem 1—we need to establish some preliminary results. We begin with the following proposition:

Proposition 1 ([8,13]). *The following results about the Zarankiewicz number are true:*

- $z(K_{17,17}, 2) = 74$.
- $z(K_{16,17}, 2) \leq 71$.
- $z(K_{17,17}, 3) \leq 141$.
- $z(K_{16,17}, 3) \leq 133$.
- $z(K_{13,17}, 3) \leq 110$.
- $z(K_{12,17}, 3) \leq 103$.
- $z(K_{11,17}, 3) \leq 96$.

Proof of Proposition 1. By using the bounds in Table 3 and Table 4 of [8] and Table C.3 of [13], the proposition holds. □

Theorem 2 ([8]). $17 \leq B(2, 2, 3) \leq 18$.

Proof of Theorem 2. The lower bound witness is found in Table 2 of [8]. The upper bound is implied by using the bounds in Table 3 and Table 4 of [8]. We know that $z(K_{18,18}, 2) = 81$, $z(K_{18,18}, 3) \leq 156$, and $2 \times 81 + 156 = 318 < 324 = |E(K_{18,18})|$. □

Suppose that (G^r, G^b, G^g) is a 3-edge coloring of $K_{17,17}$, where $K_{2,2} \nsubseteq G^r$, $K_{2,2} \nsubseteq G^b$ and $K_{3,3} \nsubseteq G^g$; in the following theorem, we specify some properties of the subgraph with color g. The properties are regarding $\Delta(G^g)$, $\delta(G^g)$, $E(G^g)$, and degree sequence of vertices X, Y in the induced graph with color g.

Theorem 3. *Assume that (G^r, G^b, G^g) is a 3-edge coloring of $K_{17,17}$, where $K_{2,2} \nsubseteq G^r$, $K_{2,2} \nsubseteq G^b$, and $K_{3,3} \nsubseteq G^g$. So:*

(a) $|E(G^g)| = 141$.
(b) $\Delta(G^g) = 9$ and $\delta(G^g) = 8$.
(c) $D_{G^g}(X) = D_{G^g}(Y) = (9, 9, 9, 9, 9, 8, 8, \ldots, 8)$.

Proof of Theorem 3. Assume that $X = \{x_1, x_2, \ldots, x_{17}\}$, $Y = \{y_1, y_2, \ldots, y_{17}\}$ is a partition set of $K = K_{17,17}$ and (G^r, G^b, G^g) is a 3-edge coloring of K, where $K_{2,2} \nsubseteq G^r$, $K_{2,2} \nsubseteq G^b$, and $K_{3,3} \nsubseteq G^g$. Since $|E(K)| = 289$, if $|E(G^g)| \leq 140$ then $|E(\overline{G^g})| \geq 149$—that is, either $|E(G^r)| \geq 75$ or $|E(G^b)| \geq 75$. In any case, by Proposition 1, either $K_{2,2} \subseteq G^r$ or $K_{2,2} \subseteq G^b$, a contradiction. Hence, assume that $|E(G^g)| \geq 141$. If $|E(G^g)| \geq 142$ then by Proposition 1, $K_{3,3} \subseteq G^g$, a contradiction again; that is, $|E(G^g)| = 141$ and part (a) is true.

To prove part (b), since $|E(G^g)| = 141$ by part (a), we can check that $\Delta(G^g) \geq 9$. Assume that there exists a vertex of $V(K)$ say x, such that $|N_{G^g}(x)| \geq 10$—that is, $\Delta(G^g) \geq 10$. Consider x and set $G_1^g = G^g \setminus \{x\}$, hence by part (a), $|E(G_1^g)| \leq 141 - 10 = 131$. Therefore, since $|E(K_{16,17})| = 272$, so $|E(\overline{G_1^g})| \geq 141$—that is, either $|E(G_1^r)| \geq 71$ or $|E(G_1^b)| \geq 71$. In any case, by Proposition 1 either $K_{2,2} \subseteq G_1^r \subseteq G^r$ or $K_{2,2} \subseteq G_1^b \subseteq G^b$, a contradiction. So, $\Delta(G^g) = 9$. To prove $\delta(G^g) = 8$, assume that $M = \{x \in X, |N_{G^g}(x)| = 9\}$ and $N = \{x \in X, |N_{G^g}(x)| = 8\}$; by part (a) one can say that $|M| \geq 5$, if $|M| = 6$, then $\delta(G^g) \leq 7$—that is, there is a vertex of X (say x) such that $|N_{G^g}(x)| \leq 7$; therefore,

$|N| \leq 10$. If $|N| = 10$, then $|E(G^g[M \cup N, Y])| = 134$, so by Proposition 1, $K_{3,3} \subseteq G^g$, a contradiction. Now assume that $|N| \leq 9$, thus $|E(G^g)| \leq (6 \times 9) + (9 \times 8) + (2 \times 7) = 140$, a contradiction again. For $|M| = 7$ if $|N| \geq 6$, then $|E(G^g[M \cup N', Y])| = 111$, where $N' \subseteq N$ and $|N'| = 6$, so by Proposition 1, $K_{3,3} \subseteq G^g$, a contradiction. Hence assume that $|N| \leq 5$; therefore, $|E(G^g)| \leq (7 \times 9) + (5 \times 8) + (5 \times 7) = 138$, a contradiction again. For $|M| = 8$ if $|N| \geq 5$, then $|E(G^g[M \cup N', Y])| = 112$, where $N' \subseteq N$ and $|N'| = 5$; therefore, by Proposition 1, $K_{3,3} \subseteq G^g$, a contradiction, so assume that $|N| \leq 4$—that is, $|E(G^g)| \leq (8 \times 9) + (4 \times 8) + (5 \times 7) = 139$, a contradiction again. For $|M| = 9$ if $|N| \geq 3$, then $|E(G^g[M \cup N', Y])| = 105$, where $N' \subseteq N$ and $|N'| = 3$, so by Proposition 1, $K_{3,3} \subseteq G^g$, a contradiction. Thus $|N| \leq 2$—that is, $|E(G^g)| \leq (9 \times 9) + (2 \times 8) + (6 \times 7) = 139$, which is a contradiction again. For $|M| = 10$, if $|N| \geq 1$, then $|E(G^g[M \cup N', Y])| = 98$, where $N' \subseteq N$ and $|N'| = 1$; so, by Proposition 1 $K_{3,3} \subseteq G^g$, a contradiction. Thus, assume that $|N| = 0$, so $|E(G^g)| \leq (10 \times 9) + (7 \times 7) = 139$, a contradiction again. Therefore, $|M| = 5$ and $|N| = 12$—that is, $\delta(G^g) = 8$, and part (b) is true.

Now, by parts (a) and (b) it is straightforward to say that $D_{G^g}(X) = D_{G^g}(Y) = (9, 9, 9, 9, 9, 8, 8, \ldots, 8)$—that is, part (c) is true, and this completes the proof. □

3. Proof of the Main Theorem

In this section, by using the results of Section 2, we will prove the main theorem.

Suppose that (G^r, G^b, G^g) is a 3-edge coloring of $K_{17,17}$, where $K_{2,2} \not\subseteq G^r$, $K_{2,2} \not\subseteq G^b$ and $K_{3,3} \not\subseteq G^g$. In the following theorem, we discuss the maximum number of common neighbors of $G^g(x)$ and $G^g(x')$ for $x, x' \in X$.

Theorem 4. *Assume that (G^r, G^b, G^g) is a 3-edge coloring of $K_{17,17}$, where $K_{2,2} \not\subseteq G^r$, $K_{2,2} \not\subseteq G^b$ and $K_{3,3} \not\subseteq G^g$. Let $|N_{G^g}(x)| = 9$ and $N_{G^g}(x) = Y_1$; the following results are true:*

(a) *For each $x \in X \setminus \{x_1\}$, we have $|N_{G^g}(x) \cap Y_1| \leq 5$.*

(b) *Assume that $n = \sum_{i=1}^{i=17} |N_{G^g}(x_i) \cap Y_1|$, then $72 \leq n \leq 73$.*

Proof of Theorem 4. Assume that $X = \{x_1, x_2, \ldots, x_{17}\}$, $Y = \{y_1, y_2, \ldots, y_{17}\}$ is a partition set of $K = K_{17,17}$, and (G^r, G^b, G^g) is a 3-edge coloring of K, where $K_{2,2} \not\subseteq G^r$, $K_{2,2} \not\subseteq G^b$ and $K_{3,3} \not\subseteq G^g$. Without loss of generality (W .l.g.) assume that $x = x_1$ and $Y_1 = \{y_1, \ldots, y_9\}$. To prove part (a), by contrast assume that there exists a vertex of $X \setminus \{x_1\}$ (say x) such that $|N_{G^g}(x) \cap Y_1| \geq 6$. W.l.g., suppose that $x = x_2$ and $Y_2 = \{y_1, y_2, \ldots, y_6\} \subseteq N_{G^g}(x_2)$. Since $K_{3,3} \not\subseteq G^g$, for each $x \in X \setminus \{x_1, x_2\}$, so $|N_{G^g}(x) \cap Y_2| \leq 2$—that is, $\sum_{i=1}^{i=17} |N_{G^g}(x_i) \cap Y_2| \leq 6 + 6 + (15 \times 2) \leq 42$. Now, since $|E(G^g[X, Y_2])| \leq 42$, one can check that there exists at least one vertex of Y_2 (say y), such that $|N_{G^g}(y)| \leq 7$, a contradiction to part (c) of Theorem 3. Hence, $|N_{G^g}(x) \cap Y_1| \leq 5$ for each $x \in X \setminus \{x_1\}$—that is, part (a) is true.

To prove part (b), if $n \leq 71$, then by part (c) of Theorem 3, it can be checked that there exists at least one vertex of Y_1 (say y), such that $|N_{G^g}(y)| \leq 7$, a contradiction. Therefore, $n \geq 72$. Assume that $n \geq 74$ and let $D_{G^g}(Y_1) = (d_1, d_2, \ldots, d_9)$. Since $\sum_{i=1}^{i=17} |N_{G^g}(x_i) \cap Y_1| \geq 74$, there exist at least two vertices of Y_1 (say y', y''), such that $|N_{G^g}(y')| = |N_{G^g}(y'')| = 9$. Since $n \geq 74$ and $|X \setminus \{x_1\}| = 16$, there exists at least one vertex of $X \setminus \{x_1\}$ (say x'), such that $|N_{G^g}(x') \cap Y_1| = 5$. W.l.g., suppose that $x' = x_2$ and $N_{G^g}(x_2) \cap Y_1 = Y_2 = \{y_1, \ldots, y_5\}$. Now we have the following claims:

Claim 1. *For each $x \in X \setminus \{x_1, x_2\}$, we have $|N_{G^g}(x) \cap Y_2| = 2$ and $D_{G^g}(Y_2) = (8, 8, 8, 8, 8)$.*

Proof of Claim 1. Since $K_{3,3} \not\subseteq G^g$ for each $x \in X \setminus \{x_1, x_2\}$, thus $|N_{G^g}(x) \cap Y_2| \leq 2$—that is, $\sum_{i=1}^{i=17} |N_{G^g}(x_i) \cap Y_2| \leq 5 + 5 + (15 \times 2) \leq 40$. Now, since $|E(G^g[X, Y_2])| \leq 40$ and $|Y_2| = 5$, if there exists a vertex of X_1)(say x'), such that $|N_{G^g}(x) \cap Y_2| \leq 1$, then $|E(G^g[X, Y_2])| \leq$

39; therefore, there exists at least one vertex of Y_2 (say y), such that $|N_{G^g}(y)| \leq 7$, a contradiction to part (c) of Theorem 3. So, $|N_{G^g}(x) \cap Y_2| = 2$ and $\sum_{y \in Y_2} |N_{G^g}(y)| = 40$, therefore by part (c) of Theorem 3 $D_{G^g}(Y_2) = (8,8,8,8,8)$, and the proof of the claim is complete. □

Claim 2. $D_{G^g}(X_1) = (5,4,4,\ldots,4)$ where $X_1 = X \setminus \{x_1\}$, in other word $|N_{G^g}(x_i) \cap Y_1| = 4$ for each $i \in \{3,4,\ldots,17\}$.

Proof of Claim 2. By contradiction, assume that there exists a vertex of $X \setminus \{x_1, x_2\}$ (say x), such that $|N_{G^g}(x) \cap Y_1| = 5$. W.l.g suppose that $x = x_3$ and $N_{G^g}(x_3) \cap Y_1 = Y_3$, now by Claim 1, $|N_{G^g}(x_3) \cap Y_2| = 2$. W.l.g., assume that $Y_3 = \{y_1, y_2, y_6, y_7, y_8\}$, thus by Claim 1, $D_{G^g}(Y_3) = (8,8,8,8,8)$—that is, $|N_{G^g}(y)| = 8$ for each $y \in Y_1 \setminus \{y_9\}$. Since $\Delta(G^g) = 9$, we can check that $n = \sum_{i=1}^{i=17} |N_{G^g}(x_i) \cap Y_1| = \sum_{i=1}^{i=9} |N_{G^g}(y_i)| \leq (8 \times 8) + 9 = 73$, a contradiction. So, $D_{G^g}(X_1) = (5,4,4,\ldots,4)$, and the proof of the claim is complete. □

Assume that $N_{G^g}(x_2) \cap Y_1 = Y_2 = \{y_1, \ldots, y_5\}$, by Claim 1 $D_{G^g}(Y_2) = (8,8,8,8,8)$. Since there exist at lest two vertices of Y_1 (say y', y''), such that $|N_{G^g}(y')| = |N_{G^g}(y'')| = 9$, thus $y', y'' \in \{y_6, y_7, y_8, y_9\}$. W.l.g., we can suppose that $y' = y_6$ and $N_{G^g}(y_6) = X_2 = \{x_1, x_3, \ldots, x_{10}\}$. By Claim 2, $|N_{G^g}(x) \cap Y_1| = 4$ and $|N_{G^g}(x) \cap Y_2| = 2$ for each $x \in X_2 \setminus \{x_1\}$—that is, $|N_{G^g}(x) \cap \{y_7, y_8, y_9\}| = 1$ for each $x \in X_2 \setminus \{x_1\}$. Since $|X_2 \setminus \{x_1\}| = 8$ and $|N_{G^g}(x) \cap \{y_7, y_8, y_9\}| = 1$, by the pigeon-hole principle, there exists a vertex of $\{y_7, y_8, y_9\}$ (say y), such that $|N_{G^g}(y) \cap X_2 \setminus \{x_1\}| \geq 3$. W.l.g., we can suppose that $y = y_7$ and $\{x_3, x_4, x_5\} \subseteq N_{G^g}(y_7) \cap X_2 \setminus \{x_1\}$. As $|Y_2| = 5$ and $|N_{G^g}(x_i) \cap Y_2| = 2$ for $i = 3, 4, 5$, there exist $i, i' \in \{3, 4, 5\}$, such that $|N_{G^g}(x_i) \cap N_{G^g}(x_{i'}) \cap Y_2| \neq 0$. W.l.g., suppose that $i = 3, i' = 4$ and $y_1 \in N_{G^g}(x_3) \cap N_{G^g}(x_4) \cap Y_2$. Therefore, $K_{3,3} \subseteq G^g[\{x_1, x_3, x_4\}, \{y_1, y_6, y_7\}]$, a contradiction. So, $n \leq 73$ and the proof of the theorem is complete. □

In part (b) of Theorem 4, we showed that $72 \leq n = \sum_{i=1}^{i=17} |N_{G^g}(x_i) \cap Y_1| \leq 73$. Now we consider these two cases independently.

3.1. The Case That n = 73

In the following theorem, we prove that in any 3-edge coloring of $K_{17,17}$ (say (G^r, G^b, G^g)), where $K_{2,2} \not\subseteq G^r$, $K_{2,2} \not\subseteq G^b$, if there exists a vertex of $V(K)$ (say x), such that $|N_{G^g}(x)| = 9$ and $\sum_{x_i \in X \setminus \{x\}} |N_{G^g}(x_i) \cap N_{G^g}(x)| = 64$, then $K_{3,3} \subseteq G^g$.

Theorem 5. Assume that (G^r, G^b, G^g) is a 3-edge coloring of $K = K_{17,17}$, such that $K_{2,2} \not\subseteq G^r$, $K_{2,2} \not\subseteq G^b$. Assume that there exists a vertex of $V(K)$ (say x), such that $|N_{G^g}(x)| = 9$. If $\sum_{i=1}^{i=17} |N_{G^g}(x_i) \cap Y_1| = 73$ where $Y_1 = N_{G^g}(x)$, then $K_{3,3} \subseteq G^g$.

Proof of Theorem 5. By contradiction, assume that $K_{3,3} \not\subseteq G^g$. Therefore, by Theorem 3 and Theorem 4, we have the following results:

(a) $|E(G^g)| = 141$.
(b) $\Delta(G^g) = 9$ and $\delta(G^g) = 8$.
(c) $D_{G^g}(X) = D_{G^g}(Y) = (9,9,9,9,9,8,8,\ldots,8)$.
(d) For each $x' \in X \setminus \{x\}$ we have $|N_{G^g}(x) \cap N_{G^g}(x')| \leq 5$.
(e) If $A = \{x \in X, |N_{G^g}(x)| = 9\}$, then $|A| = 5$ and $72 \leq \sum_{y \in N_{G^g}(x)} |N_{G^g}(y)| \leq 73$, for each $x \in A$.

Assume that $X = \{x_1, x_2, \ldots, x_{17}\}$, $Y = \{y_1, y_2, \ldots, y_{17}\}$ is the partition set of $K = K_{17,17}$, and (G^r, G^b, G^g) is a 3-edge coloring of K, where $K_{2,2} \not\subseteq G^r$, $K_{2,2} \not\subseteq G^b$ and $K_{3,3} \not\subseteq G^g$.

W.l.g., assume that $x = x_1$, $Y_1 = \{y_1, y_2, \ldots, y_9\}$, and $n = \sum_{i=1}^{i=17} |N_{G^s}(x_i) \cap Y_1| = 73$. Since $n = 73$, by (c) we can say that $D_{G^s}(Y_1) = (d_1, d_2, \ldots, d_9) = (9, 8, 8, \ldots, 8)$—that is, there exists a vertex of Y_1 (say y), such that $|N_{G^s}(y)| = 9$. By (d), $|N_{G^s}(x_1) \cap N_{G^s}(x)| \leq 5$ for each $x \in X\{\backslash x_1\}$. Set $C = \{x \in X, |N_{G^s}(x) \cap N_{G^s}(x_1)| = 5\}$. Now by argument similar to the proof of Claim 1, we have the following claim:

Claim 3. *Assume that $x \in C$ and $N_{G^s}(x) \cap Y_1 = Y'$, then for each $x' \in X \setminus \{x_1, x\}$, we have $|N_{G^s}(x') \cap Y'| = 2$ and $D_{G^s}(Y') = (8, 8, 8, 8, 8)$.*

Here there exists a claim about $|C|$ as follows:

Claim 4. $|C| \leq 2$.

Proof of Claim 4. By contradiction, assume that $|C| \geq 3$. W.l.g., suppose that $\{x_2, x_3, x_4\} \subseteq C$ and $N_{G^s}(x_2) \cap Y_1 = Y_2 = \{y_1, \ldots, y_5\}$. By Claim 3, $|N_{G^s}(x) \cap Y_2| = 2$ for each $x \in X \setminus \{x_1, x_2\}$. W.l.g., suppose that $N_{G^s}(x_3) \cap Y_1 = Y_3 = \{y_1, y_2, y_6, y_7, y_8\}$. Since $x_4 \in C$ and $|N_{G^s}(x_4) \cap Y_i| = 2$ for $i = 2, 3$, $y_9 \in N_{G^s}(x_4) \cap Y_1$. Hence, for each $y \in Y_1$, there is at least one $i \in \{2, 3, 4\}$ such that $y \in N_{G^s}(x_i)$; therefore, by Claim 3, $D_{G^s}(Y_1) = (8, 8, 8, 8, 8, 8, 8, 8, 8)$, which is in contrast to $\sum_{i=1}^{i=17} |N_{G^s}(x_i) \cap Y_1| = \sum_{i=1}^{i=9} |N_{G^s}(y_i)| = 73$, so $|C| \leq 2$. □

Now by considering $|C|$ there are three cases as follows:
Case 1: $|C| = 0$. Since $n = 73$, $|Y_1| = 9$ and $|C| = 0$, $D_{G^s}(X \setminus \{x_1\}) = (4, 4, \ldots, 4, 4)$, $D_{G^s}(Y_1) = (9, 8, 8, 8, 8, 8, 8, 8)$, $\sum_{i=1}^{i=17} |N_{G^s}(x_i) \cap Y'| = 68$, and $D_{G^s}(Y') = (9, 9, 9, 9, 8, 8, 8, 8)$, where $Y' = Y \setminus Y_1$. Set $B = \{y \in Y', |N_{G^s}(y)| = 9\}$, so $|B| = 4$.
Now we are ready to prove the following claim:

Claim 5. *There exists a vertex of $A \setminus \{x_1\}$ (say x), such that:*

$$\sum_{y \in N_{G^s}(x)} |N_{G^s}(y)| \geq 74,$$

in which $A = \{x \in X, |N_{G^s}(x)| = 9\}$.

Proof of Claim 5. $D_{G^s}(X_1) = (4, 4, \ldots, 4, 4)$ and $D_{G^s}(Y_1) = (9, 8, 8, 8, 8, 8, 8, 8)$ for each $x \in A \setminus \{x_1\}$; thus:

$$\sum_{y \in N_{G^s}(x) \cap Y_1} |N_{G^s}(y)| \geq 32$$

As $|Y'| = 8$, $|B| = 4$, and $|N_{G^s}(x_i) \cap Y'| = 5$ for each $x \in A \setminus \{x_1\}$, there exists at least one vertex of $A \setminus \{x_1\}$ (say x), such that $|N_{G^s}(x) \cap B| \geq 2$, otherwise $K_{3,3} \subseteq G^s[A, Y' \setminus B]$, a contradiction. Hence, w.l.g., suppose that $x_2 \in A$, where $|N_{G^s}(x_2) \cap B| \geq 2$. So:

$$\sum_{y \in N_{G^s}(x_2) \cap Y'} |N_{G^s}(y)| \geq 42.$$

That is,

$$\sum_{y \in N_{G^s}(x_2)} |N_{G^s}(y)| = \sum_{y \in N_{G^s}(x_2) \cap Y'} |N_{G^s}(y)| + \sum_{y \in N_{G^s}(x) \cap Y_1} |N_{G^s}(y)| \geq 42 + 32 = 74.$$

Now by considering x_2 and $N_{G^s}(x_2)$ and by (e) (or part (b) of Theorem 4) $K_{3,3} \subseteq G^s$, a contradiction again. □

Case 2: $|C| = 1$. W.l.g., suppose that $C = \{x_2\}$, $N_{G^g}(x_2) \cap Y_1 = Y_2 = \{y_1, \ldots, y_5\}$. By Claim 3, $|N_{G^g}(x_2) \cap N_{G^g}(x) \cap Y_1| = 2$ for each $x \in X \setminus \{x_1, x_2\}$ and $|N_{G^g}(y_i)| = 8$ for each $i \in \{1, 2, \ldots, 5\}$. Since there exists a vertex of Y_1 named y, such that $|N_{G^g}(y)| = 9$, w.l.g. we can suppose that $y = y_6$ and $N_{G^g}(y_6) = \{x_1, x_3, x_4 \ldots, x_{10}\}$. Since $n = 73$ and $|C| = 1$, $D_{G^g}(X_1) = (5, 4, 4, \ldots, 4, 3)$—that is, there exist at least seven vertices of $N_{G^g}(y_6) \setminus \{x_1\}$ (say $X_3 = \{x_3, x_4 \ldots, x_9\}$), such that $|N_{G^g}(x) \cap Y_1| = 4$ for each $x \in X_3$. Since $|X_3| = 7$, $|Y_2| = 5$, $|N_{G^g}(x) \cap Y_1| = 4$ and $|N_{G^g}(x_i) \cap Y_2| = 2$ for each $x \in X_3$, $|N_{G^g}(x) \cap \{y_7, y_8, y_9\}| = 1$ for each $x \in X_3$. Therefore, by the pigeon-hole principle there exists a vertex of $\{y_7, y_8, y_9\}$ (say y'), such that $|N_{G^g}(y') \cap X_3| \geq 3$. W.l.g., suppose that $y' = y_7$ and $\{x_3, x_4, x_5\} \subseteq N_{G^g}(y_7)$. Therefore, since $|Y_2| = 5$, there exists $i, i' \in \{3, 4, 5\}$ such that $|N_{G^g}(x_i) \cap N_{G^g}(x_{i'}) \cap Y_2| \neq 0$. W.l.g., suppose that $i = 3, i' = 4$ and $y_1 \in N_{G^g}(x_3) \cap N_{G^g}(x_4) \cap Y_2$. Therefore, $K_{3,3} \subseteq G^g[\{x_1, x_3, x_4\}, \{y_1, y_6, y_7\}]$, which is a contradiction.

Case 3: $|C| = 2$. W.l.g., suppose that $C = \{x_2, x_3\}$, $N_{G^g}(x_2) \cap Y_1 = Y_2 = \{y_1, \ldots, y_5\}$. By Claim 3, $|N_{G^g}(x_2) \cap N_{G^g}(x_3) \cap Y_1| = 2$. So, w.l.g. we can suppose that $N_{G^g}(x_3) \cap Y_1 = Y_3 = \{y_1, y_2, y_6, y_7, y_8\}$. Now, by Claim 3, $|N_{G^g}(y_i)| = 8$ for each $i \in \{1, 2, \ldots, 8\}$. Since there is a vertex of Y_1 named y, such that $|N_{G^g}(y)| = 9$, $y = y_9$. W.l.g., we can assume that $N_{G^g}(y_9) = X_2 = \{x_1, x_4, x_5 \ldots, x_{11}\}$. Since $n = 73$ and $|C| = 2$, $D_{G^g}(X_1) = (5, 5, 4, 4, \ldots, 4, 3, 3)$—that is, there exist two vertices of X (say x, x'), such that $|N_{G^g}(x) \cap Y_1| = 3$. If $|N_{G^g}(y_9) \cap \{x, x'\}| \leq 1$, then there exist at least seven vertices of $N_{G^g}(y_9) \setminus \{x_1\}$, such that $|N_{G^g}(x) \cap Y_1| = 4$; in this case, the proof is the same as Case 1. Hence, assume that $x, x' \in N_{G^g}(y_9)$. Since $|N_{G^g}(x) \cap Y_2| = |N_{G^g}(x') \cap Y_2| = 2$, one can check that $|N_{G^g}(x) \cap \{y_6, y_7, y_8\}| = |N_{G^g}(x') \cap \{y_6, y_7, y_8\}| = 0$. Assume that $X_i = N_{G^g}(y_i)$ for $i = 6, 7, 8$. Since $|X_i| = 8$ and $x, x' \notin X_i$, then for each $x \in X_i \setminus \{x_1\}$ we have $|N_{G^g}(x) \cap Y_1| = 4$. Therefore, by considering $X_i \setminus \{x_1\}$ and y_i for each $i \in \{6, 7, 8\}$, the proof is the same as Case 1 and $K_{3,3} \subseteq G^g$, a contradiction again.

Therefore, by Cases 1, 2, and 3 the assumption does not hold—that is, $K_{3,3} \subseteq G^g$ and this completes the proof of the theorem. □

3.2. The Case That n = 72

In the following theorem, we prove that in any 3-edge coloring of $K_{17,17}$ (say (G^r, G^b, G^g)), where $K_{2,2} \not\subseteq G^r$, $K_{2,2} \not\subseteq G^b$, if there exists a vertex of $V(K)$ (say x), such that $|N_{G^g}(x)| = 9$ and $\sum_{x_i \in X \setminus \{x\}} |N_{G^g}(x_i) \cap N_{G^g}(x)| = 63$, then $K_{3,3} \subseteq G^g$.

Theorem 6. *Assume that (G^r, G^b, G^g) is a 3-edge coloring of $K = K_{17,17}$, where $K_{2,2} \not\subseteq G^r$, $K_{2,2} \not\subseteq G^b$. Suppose that there exists a vertex of $V(K)$ (say x), such that $|N_{G^g}(x)| = 9$. If $\sum_{i=1}^{i=17} |N_{G^g}(x_i) \cap Y_1| = 72$, where $Y_1 = N_{G^g}(x)$, then $K_{3,3} \subseteq G^g$.*

Proof of Theorem 6. By contradiction, assume that $K_{3,3} \not\subseteq G^g$. Therefore, by Theorems 3 and 4, we have the following results:

(a) $|E(G^g)| = 141$.
(b) $\Delta(G^g) = 9$ and $\delta(G^g) = 8$.
(c) $D_{G^g}(X) = D_{G^g}(Y) = (9, 9, 9, 9, 9, 8, 8, \ldots, 8)$.
(d) For each $x \in X \setminus \{x_1\}$, we have $|N_{G^g}(x) \cap Y_1| \leq 5$.
(e) If $A = \{x \in X, |N_{G^g}(x)| = 9\}$, then $|A| = 5$ and $72 \leq \sum_{y \in N_{G^g}(x)} |N_{G^g}(y)| \leq 73$, for each $x \in A$.

Assume that $X = \{x_1, x_2, \ldots, x_{17}\}$, $Y = \{y_1, y_2, \ldots, y_{17}\}$ is a partition set of $K = K_{17,17}$, and (G^r, G^b, G^g) is a 3-edge coloring of K, where $K_{2,2} \not\subseteq G^r$, $K_{2,2} \not\subseteq G^b$ and $K_{3,3} \not\subseteq G^g$. W.l.g., assume that $x = x_1$, $Y_1 = \{y_1, y_2, \ldots, y_9\}$, and $n = \sum_{i=1}^{i=17} |N_{G^g}(x_i) \cap Y_1| = 72$. Since

$n = 73$, by (c) we can say that $D_{G^s}(Y_1) = (d_1, d_2, \ldots, d_9) = (8, 8, 8, \ldots, 8)$. Set $C = \{x \in X, |N_{G^s}(x) \cap N_{G^s}(x_1)| = 5\}$. Define D and E as follows:

$$D = \{x \in X \setminus \{x_1\}, \text{ such that } |N_{G^s}(x) \cap Y_1| = 5\}$$

$$E = \{x \in X \setminus \{x_1\}, \text{ such that } |N_{G^s}(x) \cap Y_1| = 3\}.$$

Here we have a claim about $|D|$ and $|E|$ as follows:

Claim 6. $|D| \leq 3$ and $|E| \leq 4$.

Proof of Claim 6. By contradiction, suppose that $|D| \geq 4$. W.l.g., assume that $\{x_2, x_3, x_4, x_5\} \subseteq D$, $N_{G^s}(x_2) \cap Y_1 = Y_2 = \{y_1, \ldots, y_5\}$. Now, by Claim 3, $|N_{G^s}(x) \cap Y_2| = 2$ for each $x \in X \setminus \{x_1, x_2\}$. W.l.g., we can suppose that $N_{G^s}(x_3) \cap Y_1 = Y_3 = \{y_1, y_2, y_6, y_7, y_8\}$. Consider $N_{G^s}(x_i) \cap Y_1 (i = 4, 5)$. Since $|N_{G^s}(x_i) \cap Y_j| = 2$ ($i = 4, 5, j = 2, 3$) and $x_i \in A$, $|N_{G^s}(x_i) \cap \{y_3, y_4, y_5\}| = 2$, $|N_{G^s}(x_i) \cap \{y_6, y_7, y_8\}| = 2$, and $y_9 \in N_{G^s}(x_i)$ for $i = 4, 5$; otherwise, if there exists a vertex of $\{x_4, x_5\}$ (say x), such that $|N_{G^s}(x_i) \cap \{y_1, y_2\}| \neq 2$, then $K_{3,3} \subseteq G^s[\{x_1, x_i, x\}, Y_1]$ for some $i \in \{1, 2\}$, a contradiction. Therefore, since $|\{y_3, y_4, y_5\}| = |\{y_6, y_7, y_8\}| = 3$ and $x_4, x_5 \in A$, by the pigeon-hole principle $|N_{G^s}(x_4) \cap N_{G^s}(x_5) \cap \{y_3, y_4, y_5\}| \geq 1$ and $|N_{G^s}(x_4) \cap N_{G^s}(x_5) \cap \{y_6, y_7, y_8\}| \geq 1$. W.l.g., we can suppose that $y_3, y_6 \in N_{G^s}(x_4) \cap N_{G^s}(x_5)$, since $y_9 \in N_{G^s}(x_4) \cap N_{G^s}(x_5)$, so $K_{3,3} \subseteq G^s[\{x_1, x_4, x_5\}, \{y_3, y_6, y_9\}]$, a contradiction. Therefore, $|D| \leq 3$. Now, as $\sum_{i=2}^{i=17} |N_{G^s}(x_i) \cap Y_1| = 63$ and $|D| \leq 3$, we can say that $|E| \leq 4$ and the proof of the claim is complete. □

Now, by considering $|D|$, there are three cases as follows:

Case 1: $|D| = 0$. Since $n = 72$ and $|D| = 0$, $D_{G^s}(X \setminus \{x_1\}) = (4, 4, \ldots, 4, 3)$, $D_{G^s}(Y_1) = (8, 8, 8, 8, 8, 8, 8, 8, 8)$, $\sum_{i=1}^{i=17} |N_{G^s}(x_i) \cap Y'| = 69$ and $D_{G^s}(Y') = (9, 9, 9, 9, 9, 8, 8, 8)$, where $Y' = Y \setminus Y_1$. Set $B = \{y \in Y', |N_{G^s}(y)| = 9\}$, hence $|B| = 5$.
Now, we have the following claim:

Claim 7. There exists a vertex of $A \setminus \{x_1\}$ (say x), such that:

$$\sum_{y \in N_{G^s}(x)} |N_{G^s}(y)| \geq 75,$$

in which $A = \{x \in X, |N_{G^s}(x)| = 9\}$.

Proof of Claim 7. Since $D_{G^s}(X_1) = (4, 4, \ldots, 4, 3)$ and $D_{G^s}(Y_1) = (8, 8, 8, 8, 8, , 8, 8, 8)$, so for at least three vertices of $A \setminus \{x_1\}$,

$$\sum_{y \in N_{G^s}(x) \cap Y_1} |N_{G^s}(y)| \geq 32.$$

Therefore, since $|N_{G^s}(x_i) \cap Y'| = 5$ for each $x \in A \setminus \{x_1\}$ and $D_{G^s}(Y') = (9, 9, 9, 9, 9, 8, 8, 8)$, there exists at least one vertex of $A \setminus \{x_1\}$ (say x), such that $|N_{G^s}(x) \cap B| \geq 3$; otherwise, $K_{3,3} \subseteq G^s[A, Y' \setminus B]$, a contradiction. Hence, w.l.g., suppose that $x_2 \in A$ and $|N_{G^s}(x_2) \cap B| \geq 3$; therefore:

$$\sum_{y \in N_{G^s}(x) \cap Y'} |N_{G^s}(y)| \geq 3 \times 9 + 2 \times 8 = 43.$$

That is, we have:

$$\sum_{y \in N_{G^s}(x_2)} |N_{G^s}(y)| = \sum_{y \in N_{G^s}(x_2) \cap Y'} |N_{G^s}(y)| + \sum_{y \in N_{G^s}(x) \cap Y_1} |N_{G^s}(y)| \geq 43 + 32 = 75.$$

Now, by considering x_2 and $N_{G^g}(x_2)$ and by (e) (or by part (b) of Theorem 4), $K_{3,3} \subseteq G^g$, a contradiction again. □

Case 2: $|D| = 1$ (for the case that $|D| = 2$, the proof is same). W.l.g., assume that $D = \{x_2\}$, $N_{G^g}(x_2) \cap Y_1 = Y_2 = \{y_1, \ldots, y_5\}$. Since $n = 72$, $|D| = 1$ and $|N_{G^g}(x) \cap Y_1| \leq 5$, $|E| = 2$. As $|N_{G^g}(x) \cap Y_2| = 2$ for each $x \in X \setminus \{x_1, x_2\}$ and $|E| = 2$, there exists a vertex of $\{y_6, y_7, y_8, y_9\}$ (say y), such that for each vertex of $N_{G^g}(y) \cap X \setminus \{x_1\}$ (say x), $|N_{G^g}(x) \cap Y_1| = 4$. W.l.g., we can suppose that $y = y_6$, $N_{G^g}(y_6) \cap X \setminus \{x_1\} = \{x_3, x_4, \ldots, x_9\}$. Since $|N_{G^g}(y_6) \cap X \setminus \{x_1\}| = 7$ and $|N_{G^g}(x) \cap Y_2| = 2$ for each $x \in N_{G^g}(y_6) \cap X \setminus \{x_1\}$, $|N_{G^g}(x) \cap \{y_7, y_8, y_9\}| = 1$. Therefore, by the pigeon-hole principle there exists a vertex of $\{y_7, y_8, y_9\}$ (say y'), such that $|N_{G^g}(y_6) \cap N_{G^g}(y') \cap X \setminus \{x_1\}| \geq 3$. W.l.g., suppose that $y' = y_7$ and $\{x_3, x_4, x_5\} \subseteq N_{G^g}(y_6) \cap N_{G^g}(y_7) \cap X \setminus \{x_1\}$. Therefore, since $|Y_2| = 5$ and $|N_{G^g}(x) \cap Y_2| = 2$, there exist at least two vertices of $\{x_3, x_4, x_5\}$ (say x', x''), such that $|N_{G^g}(x') \cap N_{G^g}(x'') \cap Y_2| \neq 0$. W.l.g., suppose that $x' = x_3, x'' = x_4$ and $y_1 \in N_{G^g}(x_3) \cap N_{G^g}(x_4)$. Therefore, $K_{3,3} \subseteq G^g[\{x_1, x_3, x_4\}, \{y_1, y_6, y_7\}]$, a contradiction.

Case 3: $|D| = 3$. W.l.g., suppose that $D = \{x_2, x_3, x_4\}$, $N_{G^g}(x_2) \cap Y_1 = Y_2 = \{y_1, \ldots, y_5\}$. By Claim 3, $|N_{G^g}(x_2) \cap N_{G^g}(x_3) \cap Y_1| = 2$. W.l.g., we can assume that $N_{G^g}(x_3) \cap Y_1 = Y_3 = \{y_1, y_2, y_6, y_7, y_8\}$. Since $x_4 \in D$ and $|N_{G^g}(x_4) \cap Y_i| = 2$ for $i = 2, 3$, $y_9 \in N_{G^g}(x_4)$. If $|N_{G^g}(x_4) \cap \{y_1, y_2\}| \neq 0$, as $|N_{G^g}(x_2) \cap N_{G^g}(x_4) \cap Y_1| = 2$ and $x_4 \in D$, one can check that $|N_{G^g}(x_4) \cap \{y_6, y_7, y_8\}| = 2$—that is, $K_{3,3} \subseteq G^g[\{x_1, x_3, x_4\}, Y_1]$, a contradiction. Hence, $|N_{G^g}(x_4) \cap \{y_1, y_2\}| = 0$. Therefore, $|N_{G^g}(x_4) \cap \{y_3, y_4, y_5\}| = 2$ and $|N_{G^g}(x_4) \cap \{y_6, y_7, y_8\}| = 2$. W.l.g., we can suppose that $N_{G^g}(x_4) \cap Y_1 = Y_4 = \{y_3, y_4, y_6, y_7, y_9\}$. Since $|D| = 3$, so $|E| = 4$. W.l.g., suppose that $E = \{x_5, x_6, x_7, x_8\}$. Here, we have a claim as follows:

Claim 8. $|N_{G^g}(y_9) \cap E| = 0$.

Proof of Claim 8. By contradiction, suppose that $|N_{G^g}(y_9) \cap E| \neq 0$. Assume that $x_5 \in N_{G^g}(y_9) \cap E$—that is, $x_5 y_9 \in E(G^g)$. Since $x_5 \in E$ and $\{x_2, x_3, x_4\} = D$, by Claim 3, $|N_{G^g}(x_5) \cap N_{G^g}(x_i)| = |N_{G^g}(x_5) \cap Y_i| = 2$ for $i = 2, 3, 4$. Consider $N_{G^g}(x_5) \cap Y_2$, assume that $N_{G^g}(x_5) \cap Y_2 = \{y', y''\}$, if $\{y', y''\} = \{y_1, y_2\}$, then $|N_{G^g}(x_5) \cap Y_4| = 1$, a contradiction. Therefore, we can assume that $|\{y', y''\} \cap \{y_1, y_2\}| \leq 1$. If $|\{y', y''\} \cap \{y_1, y_2\}| = 0$, then $|N_{G^g}(x_5) \cap Y_3| = 0$, and if $|\{y', y''\} \cap \{y_1, y_2\}| = 1$, then $|N_{G^g}(x_5) \cap Y_3| \leq 1$. In any case there exists a vertex of D (say x'), such that $|N_{G^g}(x_5) \cap N_{G^g}(x')| = 1$, a contradiction. So, the assumption does not hold and the claim is true. □

Therefore, by Claim 8, since $|N_{G^g}(y_9) \cap D| = 0$, we can say that for any vertex of $N_{G^g}(y_9) \cap X \setminus \{x_1\}$ (say x), $|N_{G^g}(x) \cap Y_1| \geq 4$; therefore, by considering Y_2 and y_9, as $|N_{G^g}(y_9) \cap X \setminus \{x_1\}| = 7$ and $|N_{G^g}(x) \cap Y_1| \geq 4$ for each $x \in N_{G^g}(y_9) \cap X \setminus \{x_1\}$, the proof is similar to Case 1, a contradiction.

Therefore, by Cases 1, 2, and 3 the assumption does not hold—that is, $K_{3,3} \subseteq G^g$ and the proof of the theorem is complete. □

Now, combining Theorems 3–6 yields the proof of Theorem 1.

4. Discussion

There are several papers in which the bipartite Ramsey numbers have been studied. In this paper, we proved the conjecture on $B(2,2,3)$, which was proposed in 2015 and states that $B(2,2,3) = 17$. We proved this conjecture by a combinatorial argument with no computer calculations. This is significant because computing the exact value of Ramsey numbers is a challenge. To approach the proof of this conjecture, we proved four theorems as follows:

1. Assume that (G^r, G^b, G^g) is a 3-edge coloring of $K_{17,17}$, where $K_{2,2} \not\subseteq G^r$, $K_{2,2} \not\subseteq G^b$ and $K_{3,3} \not\subseteq G^g$. Hence, we have:

(a) $|E(G^g)| = 141$.

(b) $\Delta(G^g) = 9$ and $\delta(G^g) = 8$.
(c) $D_{G^g}(X) = D_{G^g}(Y) = (9,9,9,9,9,8,8,\ldots,8)$.

2. Assume that (G^r, G^b, G^g) is a 3-edge coloring of $K_{17,17}$, where $K_{2,2} \nsubseteq G^r$, $K_{2,2} \nsubseteq G^b$ and $K_{3,3} \nsubseteq G^g$. Let $|N_{G^g}(x)| = 9$ and $N_{G^g}(x) = Y_1$, the following results are true:

(a) For each $x \in X \setminus \{x_1\}$, we have $|N_{G^g}(x) \cap Y_1| \le 5$.
(b) Assume that $n = \sum_{i=1}^{i=17} |N_{G^g}(x_i) \cap Y_1|$, then $72 \le n \le 73$.

3. Assume that (G^r, G^b, G^g) is a 3-edge coloring of $K = K_{17,17}$, such that $K_{2,2} \nsubseteq G^r$, $K_{2,2} \nsubseteq G^b$. Assume that there exists a vertex of $V(K)$ (say x), such that $|N_{G^g}(x)| = 9$. If $\sum_{i=1}^{i=17} |N_{G^g}(x_i) \cap Y_1| = 73$, where $Y_1 = N_{G^g}(x)$, then $K_{3,3} \subseteq G^g$.

4. Assume that (G^r, G^b, G^g) is a 3-edge coloring of $K = K_{17,17}$, where $K_{2,2} \nsubseteq G^r$, $K_{2,2} \nsubseteq G^b$. Assume that there exists a vertex of $V(K)$ (say x), such that $|N_{G^g}(x)| = 9$. If $\sum_{i=1}^{i=17} |N_{G^g}(x_i) \cap Y_1| = 72$, where $Y_1 = N_{G^g}(x)$, then $K_{3,3} \subseteq G^g$.

One might also be able to compute $B(n_1, \ldots, n_m)$ for small i, n_i like $B(2,3,3,3)$ or $B(3,3,3,3)$ in the future, using the idea of proofs laid out in this paper.

Author Contributions: Conceptualization, Y.R. and M.G.; Formal analysis, Y.R., M.G. and S.S.; Funding acquisition, S.S.; Investigation, Y.R.; Methodology, Y.R., M.G. and S.S.; Resources, Y.R.; Supervision, S.S.; Validation, S.S.; Writing(original draft), Y.R. and M.G.; Writing(review and editing), M.G. All authors have read and agreed to the published version of the manuscript.

Funding: There was no funding for this work.

Institutional Review Board Statement: Not applicable.

Informed Consent Statement: Not applicable.

Data Availability Statement: This paper focuses on pure graph theory, not involving experiments and data.

Acknowledgments: The authors would like to thank the editors and reviewers.

Conflicts of Interest: The authors declare no conflict of interest.

References

1. Erdös, P.; Rado, R. A partition calculus in set theory. *Bull. Am. Math. Soc.* **1956**, *62*, 427–489. [CrossRef]
2. Kóvari, T.; Sós, V.; Turán, P. On a problem of K. Zarankiewicz. *Colloq. Math.* **1954**, *3*, 50–57. [CrossRef]
3. Reiman, I. Über ein Problem von K. Zarankiewicz. *Acta Math. Acad. Sci. Hung.* **1958**, *9*, 269–273. [CrossRef]
4. Goddard, W.; Henning, M.A.; Oellermann, O.R. Bipartite Ramsey numbers and Zarankiewicz numbers. *Discret. Math.* **2000**, *219*, 85–95. [CrossRef]
5. Beinere, L.W.; Schwenk, A.J. On a bipartite form of the ramsey problem. In Proceedings of the Fifth British Combinatorial Conference, Aberdeen, UK, 14–18 July 1975.
6. Exoo, G. A bipartite Ramsey number. *Graphs Comb.* **1991**, *7*, 395–396. [CrossRef]
7. Hattingh, J.H.; Henning, M.A. Star-path bipartite Ramsey numbers. *Discret. Math.* **1998**, *185*, 255–258. [CrossRef]
8. Collins, A.F.; Riasanovsky, A.W.; Wallace, J.C.; Radziszowski, S. Zarankiewicz Numbers and Bipartite Ramsey Numbers. *J. Algorithms Comput.* **2016**, *47*, 63–78.
9. Dybizbański, J.; Dzido, T.; Radziszowski, S. On some Zarankiewicz numbers and bipartite Ramsey Numbers for Quadrilateral. *ARS Comb.* **2015**, *119*, 275–287.
10. Rowshan, Y.; Gholami, M.; Shateyi, S. The Size, Multipartite Ramsey Numbers for nK2 Versus Path–Path and Cycle. *Mathematics* **2021**, *9*, 764. [CrossRef]
11. Raeisi, G. Star-path and star-stripe bipartite Ramsey numbers in multicoloring. *Trans. Comb.* **2015**, *4*, 37–42.
12. Hatala, I.; Héger, T.; Mattheus, S. New values for the bipartite Ramsey number of the four-cycle versus stars. *Discret. Math.* **2021**, *344*, 112320. [CrossRef]
13. Collins, A.F. *Bipartite Ramsey Numbers and Zarankiewicz Numbers*; Rochester Institute of Technology: Rochester, NY, USA, 2015.
14. Gholami, M.; Rowshan, Y. The bipartite Ramsey numbers $BR(C\_8, C\_\{2n\})$. *arXiv* **2021**, arXiv:2108.02630.
15. Bondy, J.A.; Murty, U.S.R. *Graph Theory with Applications*; Macmillan: London, UK, 1976; Volume 290.

Article

Total Coloring of Dumbbell Maximal Planar Graphs

Yangyang Zhou [1,2,*], Dongyang Zhao [1,2], Mingyuan Ma [1,2] and Jin Xu [1,2]

1 School of Electronics Engineering and Computer Science, Peking University, Beijing 100871, China; zdy_macs@pku.edu.cn (D.Z.); mamingyuan@pku.edu.cn (M.M.); jxu@pku.edu.cn (J.X.)
2 Key Laboratory of High Confidence Software Technologies, Peking University, Beijing 100871, China
* Correspondence: zyy_eecs@pku.edu.cn

Abstract: The Total Coloring Conjecture (TCC) states that every simple graph G is totally $(\Delta + 2)$-colorable, where Δ denotes the maximum degree of G. In this paper, we prove that TCC holds for dumbbell maximal planar graphs. Especially, we divide the dumbbell maximal planar graphs into three categories according to the maximum degree: J_9, I-dumbbell maximal planar graphs and II-dumbbell maximal planar graphs. We give the necessary and sufficient condition for I-dumbbell maximal planar graphs, and prove that any I-dumbbell maximal planar graph is totally 8-colorable. Moreover, a linear time algorithm is proposed to compute a total $(\Delta + 2)$-coloring for any I-dumbbell maximal planar graph.

Keywords: total coloring; dumbbell maximal planar graphs; I-dumbbell maximal planar graphs; dumbbell transformation; total coloring algorithm

Citation: Zhou, Y.; Zhao, D.; Ma, M.; Xu, J. Total Coloring of Dumbbell Maximal Planar Graphs. *Mathematics* 2022, 10, 912. https://doi.org/10.3390/math10060912

Academic Editors: Janez Žerovnik and Darja Rupnik Poklukar

Received: 25 February 2022
Accepted: 11 March 2022
Published: 13 March 2022

Publisher's Note: MDPI stays neutral with regard to jurisdictional claims in published maps and institutional affiliations.

Copyright: © 2022 by the authors. Licensee MDPI, Basel, Switzerland. This article is an open access article distributed under the terms and conditions of the Creative Commons Attribution (CC BY) license (https://creativecommons.org/licenses/by/4.0/).

1. Introduction

All graphs considered in this paper are simple, finite and undirected, and we follow [1] for the terminologies and notations not defined here. For any graph G, we denote by $V(G)$, $E(G)$, $\Delta(G)$ and $\delta(G)$ (or simply V, E, Δ and δ) the vertex set, the edge set, the maximum degree and the minimum degree of G, respectively. If $uv \in E(G)$, then u is said to be a neighbor of v. We use $N(v)$ to denote the set of neighbors of v. The degree of v, denoted by $d(v)$, is the number of neighbors of v, i.e., $d(v) = |N(v)|$. A k-vertex is a vertex of degree k. Given a set $X \subseteq V$, we denote by $G[X]$ the subgraph of G induced by X. A k-cycle is a cycle of length k, and a 3-cycle is usually called a triangle. We use K_n to denote the complete graph of order n. For a disjoint union of G and H, the joining of G and H, denoted by $G \vee H$, is the graph obtained by joining every vertex of G to every vertex of H. The joined $C_n \vee K_1$ of a cycle and a single vertex is a wheel with n spokes, denoted by W_n, where C_n and K_1 are called the cycle and center of W_n, respectively.

A total k-coloring of G is a mapping $\phi : V \cup E \rightarrow \{1, 2, \cdots, k\}$ such that $\phi(x) \neq \phi(y)$ is for any two adjacent or incident elements $x, y \in V \cup E$. A graph G is totally k-colorable if it admits a total k-coloring. The total chromatic number $\chi''(G)$ is the smallest integer k, such that G has a total k-coloring. Behzad [2] and Vizing [3] posed independently the following famous conjecture, known as the Total Coloring Conjecture (TCC).

Conjecture 1. *For any graph* G, $\Delta(G) + 1 \leq \chi''(G) \leq \Delta(G) + 2$.

Obviously, the lower bound is trivial. The upper bound is still unproved. To date, TCC has been confirmed for general graphs with $\Delta \leq 5$ [4–7] and for planar graphs with $\Delta \geq 7$ [8–11]. Therefore, for planar graphs, the only open case is $\Delta = 6$. Nevertheless, scholars have studied the total coloring of planar graphs under some restricted conditions [12–17]. Among these, Sun et al. [13] proved that TCC is true for planar graphs without adjacent triangles. Here, adjacent triangles are two triangles that share a common edge. Zhu and Xu [17] gave a stronger statement that TCC holds for planar graphs G with

$\Delta(G) = 6$, if G does not contain any subgraph isomorphic to a 4-fan. Regardless of the results in [13] or in [17], the graph G cannot contain adjacent triangles. This leads us to study the total coloring and total chromatic number of maximal planar graphs, whose faces are all triangles. In [18], we study the total coloring of recursive maximal planar graphs and prove that TCC is true for recursive maximal planar graphs. Moreover, (2,2)-recursive maximal planar graphs are totally $(\Delta + 1)$-colorable.

A maximal planar graph is a planar graph to which no edges can be added without violating the planarity. Let G be a maximal planar graph and C be a cycle of G with $|C| \geq 4$. We call the subgraph of G induced by the vertices on C and the vertices located inside (or outside) C a semi-maximal planar graph based on C, which is denoted by G_{in}^C (or G_{out}^C). In fact, a semi-maximal planar graph is a triangulated disc.

According to the vertex coloring, maximal planar graphs can be partitioned into three categories: purely tree-colorable, purely cycle-colorable and impure colorable, refer to [19]. In [20], Xu et al. proposed the purely tree-colorable graphs conjecture, which states that a maximal planar graph is purely tree-colorable if and only if it is the icosahedron or a dumbbell maximal planar graph. They further studied the structures and properties of dumbbell maximal planar graphs in [19]. Then, what is the total coloring of dumbbell maximal planar graphs? This problem has aroused our concern.

We aim to study the total coloring of dumbbell maximal planar graphs in this paper. The remainder of this paper is organized as follows. In Section 2, we introduce the dumbbell transformation and study the structures and properties of dumbbell maximal planar graphs. In particular, we classify the dumbbell maximal planar graphs into three categories. In Section 3, we prove that any dumbbell maximal planar graph is totally $(\Delta + 2)$-colorable. In Section 4, we propose an algorithm with linear time complexity to compute a total $(\Delta + 2)$-coloring for any I-dumbbell maximal planar graph. In Section 5, we make a conclusion for the paper.

2. Dumbbell Maximal Planar Graphs

We study the structures and properties of dumbbell maximal planar graphs in this section. Before this, we need to introduce the dumbbell transformation given by Xu [19].

2.1. Dumbbell Transformation

In order to give the dumbbell transformation, we introduce the extending 3-wheel and 4-wheel operations first.

The extending 3-wheel operation. The extending 3-wheel operation acts on a triangle of a maximal planar graph, specifically, adding a new vertex in the face and joining it to every vertex of the triangular face, as shown in Figure 1.

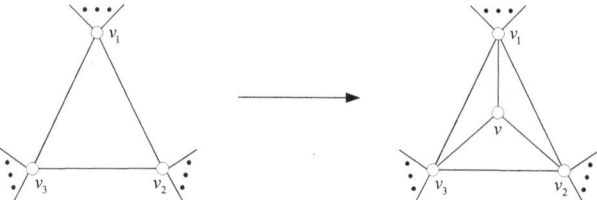

Figure 1. The extending 3-wheel operation.

The extending 4-wheel operation. The object of the extending 4-wheel operation is a path of length 2. Specifically, an extending 4-wheel operation based on path $v_1v_2v_3$ means: split the vertex v_2 into v_2 and v_2', and split the edges v_1v_2 and v_2v_3 into v_1v_2, v_1v_2' and $v_2v_3, v_2'v_3$, respectively. Hence, the vertices v_1, v_2', v_3 and v_2 form a cycle of length 4. Then, add a new vertex x in this cycle and make x adjacent to vertices v_1, v_2', v_3 and v_2, respectively. The process is shown in Figure 2.

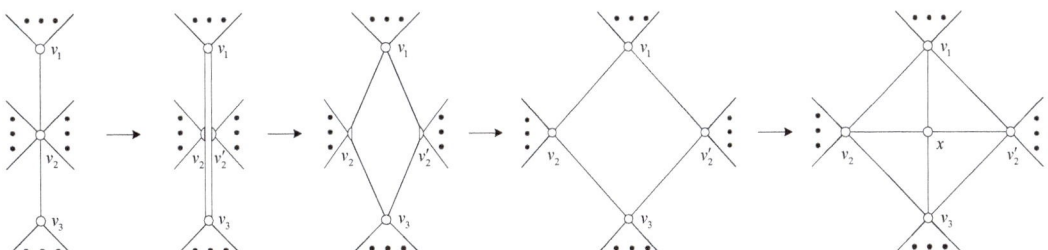

Figure 2. The extending 4-wheel operation.

A dumbbell is a graph consisting of two triangles $\triangle v_1 v_2 u$ and $\triangle u v_3 v_4$ with exactly one common vertex u, and it is denoted by $X = \triangle v_1 v_2 u \cup \triangle u v_3 v_4$, as shown in the left of Figure 3. Obviously, a 4-wheel contains exactly two dumbbells, as shown in the right of Figure 3, where $X_1 = \triangle v_1 v_2 u \cup \triangle u v_3 v_4$ and $X_2 = \triangle v_1 v_3 u \cup \triangle u v_2 v_4$. In this paper, dumbbells considered are ones contained in a 4-wheel without special assertion.

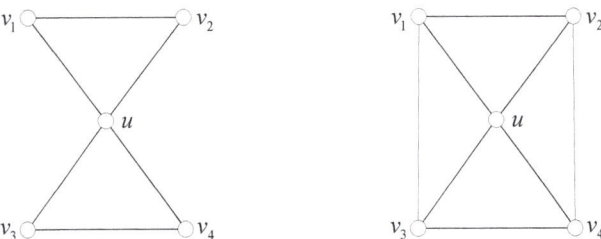

Figure 3. The dumbbell and a 4-wheel.

The dumbbell transformation. For a given dumbbell $X = \triangle v_1 v_2 u \cup \triangle u v_3 v_4$. First, add two 3-vertices x_1 and x_2 on the two triangular faces of X, respectively. Then, implement the extending 4-wheel operation on path $x_1 u x_2$, the newly added 4-vertex is denoted by v, as shown in Figure 4.

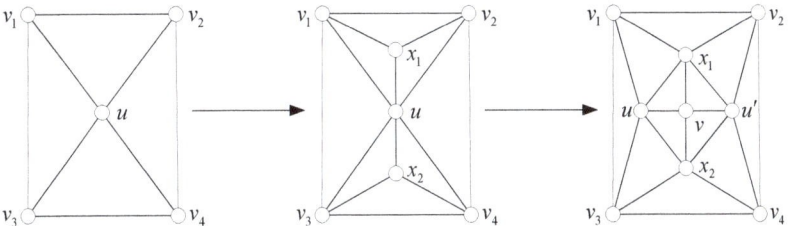

Figure 4. The dumbbell transformation.

Xu et al. [20] gave the following theorem:

Theorem 1. *Let G be a maximal planar graph with a 4-wheel W_4. Then the graphs obtained from G by implementing the dumbbell transformations on two dumbbells of W_4 are isomorphic.*

2.2. Structure and Property of Dumbbell Maximal Planar Graphs

The first maximal planar graph with order 9, denoted by J_9, is shown in Figure 5 and is called a dumbbell maximal planar graph, which is the dumbbell maximal planar graph with the minimum order. A graph is a dumbbell maximal planar graph if one of the following conditions is satisfied: (1) it is isomorphic to J_9; (2) it can be obtained from another dumbbell maximal planar graph by the dumbbell transformation. In general, if $J_{4k+1}(k \geq 2)$ is a dumbbell maximal planar graph, we call the maximal planar graph obtained from J_{4k+1} by implementing a dumbbell transformation a dumbbell maximal planar graph. Implement the dumbbell transformation on each unidentical 4-wheel in J_{4k+1}, then we can obtain dumbbell maximal planar graphs with order $4k + 5$. As shown in Figure 5, we give the dumbbell maximal planar graphs with orders 9, 13, 17 and 21, respectively.

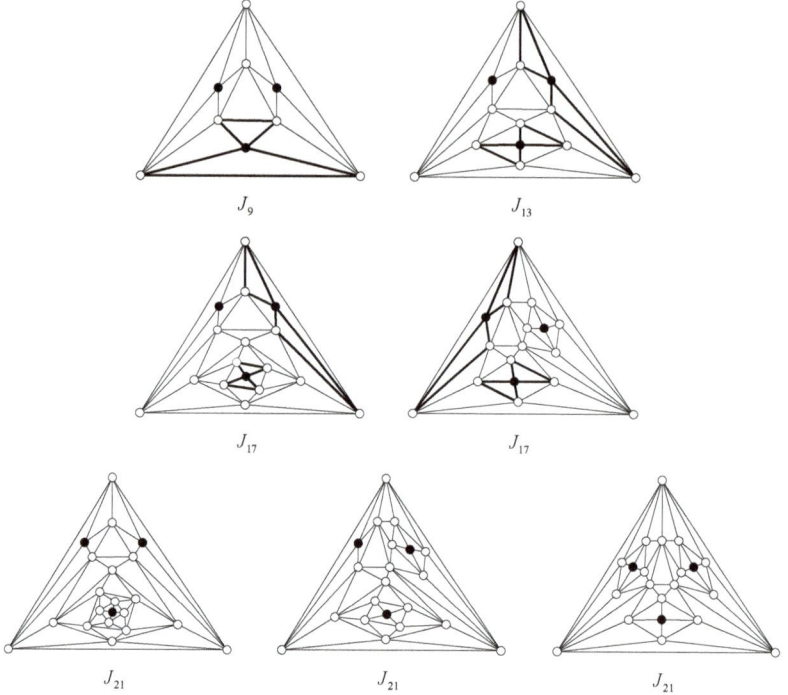

Figure 5. The dumbbell maximal planar graphs with orders 9, 13, 17 and 21.

J_9 contains exactly three vertices of degree 4. By the definition of dumbbell maximal planar graphs, Xu et al. [20] obtained the following observation.

Observation 1. *(1) Any dumbbell maximal planar graph has order $4k + 1$, where $k \geq 2$; (2) Any dumbbell maximal planar graph contains exactly three vertices of degree 4.*

We give the following theorem on the maximum degrees of dumbbell maximal planar graphs.

Theorem 2. *Except for J_9, the maximum degree of a dumbbell maximal planar graph $J_{4k+1}(k \geq 3)$ is 6 or 7.*

Proof. Obviously, the maximum degree of J_9 is 5. As shown in Figure 4, for each dumbbell transformation, the degree of each vertex on the cycle of the original 4-wheel is increased by 1, and that of the new 4-wheel is 5. As shown in Figure 5, the maximum degree of J_{13} is 6; the two non-isomorphic dumbbell maximal planar graphs J_{17}, which are obtained from J_{13} by implementing the dumbbell transformation on the two unidentical 4-wheels, have the maximum degrees 6 and 7, respectively; the three dumbbell planar graphs J_{21} obtained from J_{17} have the maximum degree 6, 7 and 7, respectively. It is observed that the degrees of vertices on the wheels of all 4-wheels in these three dumbbell maximal planar graphs with order 21 are 5 and 6, and the maximum degree of each dumbbell maximal planar graph obtained by implementing the dumbbell transformation does not exceed 7. By analogy, the maximum degree of a dumbbell maximal planar graph with higher order is always 6 or 7. □

For the dumbbell maximal planar graph with maximum degree 6, we have

Theorem 3. *The maximum degree of a dumbbell maximal planar graph G is 6 if and only if G is obtained from J_9 by continuously implementing the dumbbell transformation, and each transformation is implemented on the new 4-wheel generated by the previous transformation (The first dumbbell transformation is implemented on an arbitrary 4-wheel in J_9).*

The proof of Theorem 3 is obvious and therefore omitted.

We call the dumbbell maximal planar graphs with maximum degree 6 described in Theorem 3 I-dumbbell maximal planar graphs (The I-dumbbell maximalplanar graphs we define here are dumbbell maximal planar graphs of maximum degree 6, so of course J_9 is not included) and dumbbell maximal planar graphs with maximum degree 7 II-dumbbell maximal planar graphs. For I-dumbbell maximal planar graphs, we obtain the following observation.

Observation 2. *For any I-dumbbell maximal planar graph, the degrees of vertices on the cycle of the newly generated 4-wheel are all 5. Furthermore, the other two 4-wheels do not have this property.*

Figure 6 shows the generation process of I-dumbbell maximal planar graphs.

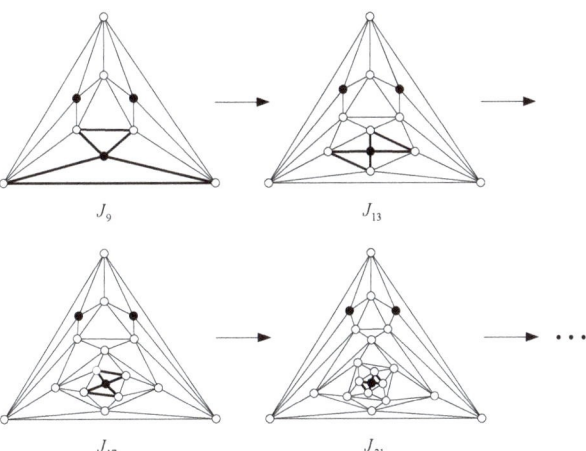

Figure 6. The schematic diagram of the generation process of I-dumbbell maximal planar graphs.

3. Total Coloring of Dumbbell Maximal Planar Graphs

In Section 2, the dumbbell transformation and dumbbell maximal planar graphs were introduced. In this section, we study the total coloring of dumbbell maximal planar graphs based on structural characteristics.

From the previous section, we know that any dumbbell maximal planar graph has exactly 3 vertices of degree 4, and the maximum degree of a dumbbell maximal planar graph is 6 or 7, except for J_9. Furthermore, we draw an important conclusion about the structure of the dumbbell maximal planar graphs.

According to the maximum degree, dumbbell maximal planar graphs can be divided into the following three categories: J_9, I-dumbbell maximal planar graphs and II-dumbbell maximal planar graphs. In Figure 7, we give a total 7-coloring of J_9. Sanders and Zhao [11] proved that planar graphs with $\Delta = 7$ are totally 9-colorable. Therefore, we only need to consider dumbbell maximal planar graphs with maximum degree 6, that is, I-dumbbell maximal planar graphs.

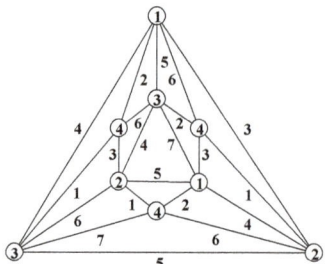

Figure 7. A total 7-coloring of J_9.

Theorem 4. *Any I-dumbbell maximal planar graph is totally 8-colorable.*

Proof. J_{13} is the I-dumbbell maximal planar graph with the minimum order, and J_{13} is totally 8-colorable, as shown in Figure 8.

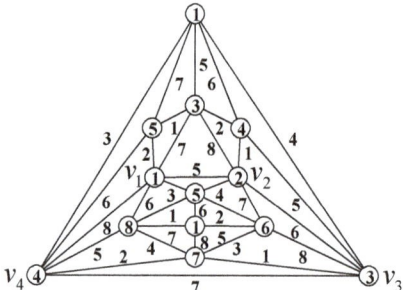

Figure 8. A total 8-coloring of J_{13}.

Since the I-dumbbell maximal planar graphs are obtained from J_9 by continuously implementing the dumbbell transformation at a unique 4-wheel only, without loss of generality, we assume that all I-dumbbell maximal planar graphs are obtained by implementing the dumbbell transformation at the 4-wheel located at the bottom of J_9, as shown in Figure 6. For convenience, we denote the cycle of the 4-wheel located at the bottom of J_9 by $C_4 = v_1 v_2 v_3 v_4 v_1$. Then, for any I-dumbbell maximal planar graph G, $G = G_{in}^{C_4} \cup G_{out}^{C_4}$, where $G_{in}^{C_4}$ and $G_{out}^{C_4}$ are the two semi-maximal planar graphs based on C_4. In the following, we give a total coloring scheme of any I-dumbbell maximal planar graph G. We color $G_{out}^{C_4}$ the same way in J_{13} and color $G_{in}^{C_4}$ according to the parity of the number of dumbbell transformations, which is denoted by l.

When l is odd, the coloring scheme is:

The colors of vertices v_1, v_2, v_3, v_4 on the cycle of the initial 4-wheel are 1, 2, 3 and 4, and the colors of the edges are 5, 6, 7 and 8, respectively;

After the first dumbbell transformation, starting from the vertex opposite the edge colored with 5, color the vertices on the cycle of newly generated 4-wheel with 5, 6, 7 and 8, and the corresponding edges with 1, 2, 3 and 4 in clockwise order;

The colors of edges between the newly generated 4-wheel and the initial 4-wheel are 3, 4, 7, 8, 1, 2, 5 and 6 in clockwise order;

After the second dumbbell transformation, starting from the vertex opposite the edge colored with 1, color the vertices on the cycle of newly generated 4-wheel with 1, 2, 3 and 4, and the corresponding edges with 5, 6, 7 and 8, in clockwise order;

The colors of the edges between the newly generated 4-wheel and the previous 4-wheel are 7, 8, 1, 4, 5, 6, 3 and 2 in clockwise order;

After the i-th ($3 \leq i \leq l$) dumbbell transformation, the colors of vertices and edges on the cycle of the newly generated 4-wheel, and the colors of edges between the newly generated 4-wheel and the previous 4-wheel, are the same as the first dumbbell transformation when i is odd; the colors of vertices and edges on the cycle of the newly generated 4-wheel, and the colors of edges between the newly generated 4-wheel and the previous 4-wheel, are the same as the second dumbbell transformation when i is even;

After the last dumbbell transformation, we specify the color of the wheel center as 1, and the colors of the spokes as 6, 5, 8 and 7 from whose end point is colored with 5 in clockwise order. Of course, the readers can also use other appropriate colors;

When l is even, the coloring scheme is similar to that when l is odd, except that the color of the wheel center is 5, and the colors of the spokes are 3, 4, 2 and 1 from whose end point is colored with 1 in clockwise order;

So, we obtain a total 8-coloring scheme of any I-dumbbell maximal planar graph, and the proof is completed. □

As shown in Figure 9, we give the coloring scheme for $l = 3$ (on the left) and $l = 4$ (on the right), respectively.

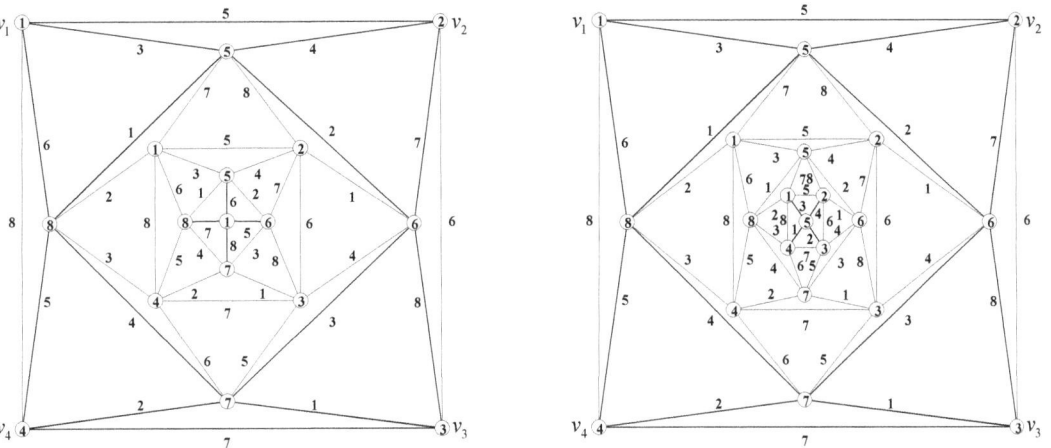

Figure 9. The coloring diagram for $l = 3$ and $l = 4$.

Therefore, we obtain the following theorem.

Theorem 5. *The TCC holds for dumbbell maximal planar graphs.*

4. Total Coloring Algorithm for I-Dumbbell Maximal Planar Graphs

In this section an algorithm with linear time complexity is proposed, which computes a total $(\Delta + 2)$-coloring for any I-dumbbell maximal planar graph. It is known that an arbitrary I-dumbbell maximal planar graph can be obtained from J_{13} by continuously implementing the dumbbell transformation on the newly generated 4-wheel. We introduce the concept of dumbbell-recursive generation sequence to formalize the generation process.

Definition 1 (Dumbbell-Recursive Generation Sequence). *Let J_{4l+13} ($l \geq 0$) be an I-dumbbell maximal planar graph with $W_4^l \triangleq J_{4l+13}[\{v_1^l, v_2^l, v_3^l, v_4^l, v^l\}]$ as the newly generated 4-wheel, where*

v_1^l, v_2^l, v_3^l, and v_4^l denote vertices on the cycle and v^l denotes the wheel center, respectively. Starting from J_{13}, each time we implement the dumbbell transformation, an I-dumbbell maximal planar graph J_{4i+13} is obtained, where $i = 1, 2, \cdots, l$. Then, the dumbbell-recursive generation sequence of J_{4l+13} is defined as $\Phi(J_{4l+13}) = \{J_{13}; W_4^0, W_4^1, \cdots, W_4^l\}$.

Now, we give a total coloring algorithm for I-dumbbell maximal planar graphs, as shown in the following Algorithm 1, which consists of two stages.

Algorithm 1 Total Coloring Algorithm for I-dumbbell Maximal Planar Graph

1: **Input:** An I-dumbbell maximal planar graph J_{4l+13}.
2: **Output:** The total coloring dictionary U.
3: **Stage 1. Dumbbell-recursive generation sequence generation.**
4:    $\Phi \leftarrow$ empty list, $i \leftarrow l$.
5: **while** i is not 0 **do**
6:      Choose the newly generated 4-wheel W_4^i according to Observation 2.
7:      Implement the inverse process of dumbbell transformation and obtain $J_{4(i-1)+13}$.
8:      Store W_4^i to Φ.
9:      $i \leftarrow i - 1$.
10: **end while**
11: Store W_4^0 and J_{13} to Φ.
12: $\Phi \leftarrow \text{reverse}(\Phi)$.
13: **Stage 2. Total coloring based on Φ.**
14: Take out J_{13} from Φ.
15: Color J_{13} as shown in Figure 8, and store the coloring information to U.
16: $i \leftarrow 0$.
17: **while** Φ is not empty **do**
18:      Take out the first element W_4^i from Φ.
19:      Implement dumbbell transformation on W_4^i and obtain $J_{4(i+1)+13}$, W_4^{i+1}.
20:      **if** $(i+1)$ is odd **then**
21:          $U[v^{i+1}] \leftarrow 1$.
22:          Color other vertices of W_4^{i+1} and the associated edges according to Theorem 4.
23:          Store the coloring information of W_4^{i+1} to U.
24:      **else**
25:          $U[v^{i+1}] \leftarrow 5$.
26:          Color other vertices of W_4^{i+1} and the associated edges according to Theorem 4.
27:          Store the coloring information of W_4^{i+1} to U.
28:      **end if**
29:      $i \leftarrow i + 1$.
30: **end while**
31: **return** U.

In the first stage, given an arbitrary I-dumbbell maximal planar graph J_{4l+13}, we compute the dumbbell-recursive generation sequence $\Phi(J_{4l+13})$. As mentioned in Observation 2, we can easily find the newly generated 4-wheel according to the degrees of vertices on the cycle. Then, the inverse process of dumbbell transformation is implemented to obtain the previous dumbbell maximal planar graph. By repeating the procedure and storing the structure information, we obtain the dumbbell-recursive generation sequence.

In the second stage, we give a total $(\Delta + 2)$-coloring of J_{4l+13} based on the dumbbell-recursive generation sequence $\Phi(J_{4l+13})$. More precisely, for J_{4l+13} with $V(J_{4l+13}) = \{v_1, v_2, \cdots, v_n\}$ and $E(J_{4l+13}) = \{e_1, e_2, \cdots, e_m\}$, let $C(J_{4l+13}) = \{1, 2, \cdots, \Delta(J_{4l+13}) + 2\}$ be the color set.

The dictionary structure $U = \{v_1 : \phi(v_1), \cdots, v_n : \phi(v_n), e_1 : \phi(e_1), \cdots, e_m : \phi(e_m)\}$ is used to store the total coloring scheme, where $\phi(v_i), \phi(e_j) \in C(J_{4l+13})$, $i = 1, \cdots, n$, $j = 1, \cdots, m$. Firstly, take out the initial graph J_{13} and color its vertices and edges as shown in Figure 8, and store the corresponding coloring information in U. Then, take out the generation operation information W_4^i stored in Φ in turn. The dumbbell transformation is implemented and the coloring information of the newly generated W_4^{i+1} is stored to U according to l's parity. Finally, a total $(\Delta + 2)$-coloring of any I-dumbbell maximal planar graph can be obtained iteratively.

During the execution of Stage 1 and Stage 2, the order of J_{4i+13} varies by 4 at each step. Furthermore, the number of sequence generation and coloring operations is constant at each step. Therefore, the time complexity of this algorithm is linear.

5. Conclusions

Total coloring is an important and representative problem in the field of graph coloring. Even for planar graphs, the total coloring conjecture is still open for the case $\Delta = 6$. In this paper, we prove that the Total Coloring Conjecture holds for dumbbell maximal planar graphs, which are generated by implementing the dumbbell transformation continuously. According to the maximum degree, we divide the dumbbell maximal planar graphs into three categories: J_9, I-dumbbell maximal planar graphs and II-dumbbell maximal planar graphs. Furthermore, we give the necessary and sufficient condition for I-dumbbell maximal planar graphs and prove that any I-dumbbell maximal planar graph is totally 8-colorable. Moreover, an algorithm with linear time complexity is presented to compute a total $(\Delta + 2)$-coloring of any I-dumbbell maximal planar graph. For future work, we will further focus on the relationship between the structure and coloring of dumbbell maximal planar graphs and discuss the condition in which the dumbbell maximal planar graphs are totally $(\Delta + 1)$-colorable.

Author Contributions: Creation and conceptualization of the idea, Y.Z. and J.X.; writing—original draft preparation, Y.Z. and D.Z.; writing—review and editing, Y.Z. and M.M. All authors have read and agreed to the published version of the manuscript.

Funding: This research was supported by the National Key R&D Program of China No. 2019YFA0706 401 and the National Natural Science Foundation of China No. 62172014, No. 62172015, No. 61872166 and No. 62002002.

Institutional Review Board Statement: Not applicable.

Informed Consent Statement: Not applicable.

Data Availability Statement: Not applicable.

Conflicts of Interest: The authors declare no conflict of interest.

References

1. Bondy, J.A.; Murty, U.S.R. *Graph Theory with Applications*; MacMillan: London, UK, 1976.
2. Behzad, M. Graphs and Their Chromatic Numbers. Ph.D. Thesis, Michigan State University, East Lansing, MI, USA, 1965.
3. Vizing, V.G. On an estimate of the chromatic class of a p-graph. *Discret. Anal.* **1964**, *3*, 25–30.
4. Rosenfeld, M. On the total coloring of certain graphs. *Isr. J. Math.* **1971**, *9*, 396–402. [CrossRef]
5. Vijayaditya, N. On total chromatic number of a graph. *J. Lond. Math. Soc.* **1971**, *2*, 405–408. [CrossRef]
6. Kostochka, A.V. The total coloring of a multigraph with maximal degree 4. *Discret. Math.* **1977**, *17*, 161–163. [CrossRef]
7. Kostochka, A.V. The total chromatic number of any multigraph with maximum degree five is at most seven. *Discret. Math.* **1996**, *162*, 199–214. [CrossRef]
8. Borodin, O.V. On the total coloring of planar graphs. *J. Reine Angew. Math.* **1989**, *394*, 180–185.
9. Jensen, T.R.; Toft, B. *Graph Coloring Problems*; Wiley: Hoboken, NJ, USA, 1995.
10. Yap, H.P. *Total Colourings of Graphs*; Lecture Notes in Mathematics; Springer: Berlin/Heidelberg, Germany, 1996.
11. Sanders, D.P.; Zhao, Y. On total 9-coloring planar graphs of maximum degree seven. *J. Graph Theory* **1999**, *31*, 67–73. [CrossRef]
12. Wang, Y.Q.; Shangguan, M.L.; Li, Q. On total chromatic number of planar graphs without 4-cycles. *Sci. China* **2007**, *50*, 81–86. [CrossRef]
13. Sun, X.Y.; Wu, J.L.; Wu, Y.W.; Hou, J.F. Total colorings of planar graphs without adjacent triangles. *Discret. Math.* **2009**, *309*, 202–206. [CrossRef]
14. Roussel, N. Local condition for planar graphs of maximum degree 6 to be total 8-colorable. *Taiwan. J. Math.* **2011**, *15*, 87–99. [CrossRef]

15. Hou, J.F.; Liu, B.; Liu, G.Z.; Wu, J.L. Total coloring of planar graphs without 6-cycles. *Discret. Appl. Math.* **2011**, *159*, 157–163. [CrossRef]
16. Wu, Q.; Lu, Q.L.; Wang, Y.Q. ($\Delta + 1$)-total-colorability of plane graphs of maximum degree ≥ 6 with neither chordal 5-cycle nor chordal 6-cycle. *Inform. Process. Lett.* **2011**, *111*, 767–772. [CrossRef]
17. Zhu, E.Q.; Xu, J. A sufficient condition for planar graphs with maximum degree 6 to be totally 8-colorable. *Discret. Appl. Math.* **2017**, *223*, 148–153. [CrossRef]
18. Zhou, Y.Y.; Zhao, D.Y.; Ma, M.Y.; Xu, J. Total coloring of recursive maximal planar graphs. *Theor. Comput. Sci.* **2022**, *909*, 12–18. [CrossRef]
19. Xu, J. Theory on structure and coloring of maximal planar graphs (3): purely tree-colorable and uniquely 4-colorable maximal planar graph conjectures. *J. Electron. Inf. Technol.* **2016**, *38*, 1328–1363.
20. Xu, J.; Li, Z.P.; Zhu, E.Q. On purely tree-colorable planar graphs. *Inform. Process. Lett.* **2016**, *116*, 532–536. [CrossRef]

Article
Domination Coloring of Graphs

Yangyang Zhou [1,2], Dongyang Zhao [1,2], Mingyuan Ma [1,2,*] and Jin Xu [1,2]

[1] School of Electronics Engineering and Computer Science, Peking University, Beijing 100871, China; zyy_eecs@pku.edu.cn (Y.Z.); zdy_macs@pku.edu.cn (D.Z.); jxu@pku.edu.cn (J.X.)
[2] Key Laboratory of High Confidence Software Technologies, Peking University, Beijing 100871, China
* Correspondence: mamingyuan@pku.edu.cn

Abstract: A domination coloring of a graph G is a proper vertex coloring of G, such that each vertex of G dominates at least one color class (possibly its own class), and each color class is dominated by at least one vertex. The minimum number of colors among all domination colorings is called the domination chromatic number, denoted by $\chi_{dd}(G)$. In this paper, we study the complexity of the k-domination coloring problem by proving its NP-completeness for arbitrary graphs. We give basic results and properties of $\chi_{dd}(G)$, including the bounds and characterization results, and further research $\chi_{dd}(G)$ of some special classes of graphs, such as the split graphs, the generalized Petersen graphs, corona products, and edge corona products. Several results on graphs with $\chi_{dd}(G) = \chi(G)$ are presented. Moreover, an application of domination colorings in social networks is proposed.

Keywords: domination coloring; domination chromatic number; split graphs; generalized Petersen graphs; corona products; edge corona products

MSC: 05C15; 05C69

1. Introduction and Preliminary

1.1. Introduction

Coloring and domination are two important fields in graph theory, and both have rich research results. For comprehensive results of coloring and domination in graphs, refer to [1–17], respectively. Moreover, graph coloring and domination problems are often in relation. Chellali and Volkmann [18] showed some relations between the chromatic number and some domination parameters in a graph. For a graph $G = (V, E)$, a vertex $v \in V$ dominates a set $S \subseteq V$ if it is adjacent to every vertex of S, meanwhile, we say that v is a dominator of S, and S is dominated by v. Hedetniemi et al. [19] introduced the concept of a dominator partition of a graph. A dominator partition is a partition $\pi = \{V_1, V_2, \cdots, V_k\}$ of $V(G)$, such that every vertex $v \in V$ is a dominator of at least one block V_i of π. Motivated by [19], Gera et al. [20] proposed the dominator coloring in 2006.

Definition 1 ([20]). *A dominator coloring of a graph G is a proper coloring, such that every vertex of G dominates at least one color class (possibly its own class). The dominator chromatic number of G, denoted by $\chi_d(G)$, is the minimum number of colors among all dominator colorings of G.*

Gera researched further in [21,22]. More results on the dominator coloring can be found in [23–26]. Kazemi [27] proposed the concept of total dominator coloring in 2015, which is a proper coloring, such that each vertex of the graph is adjacent to every vertex of some (other) color class. For more results on the total dominator coloring, refer to [28–30]. In 2015, Merouane et al. [31] proposed the dominated coloring:

Definition 2 ([31]). *A dominated coloring of a graph G is a proper coloring such that every color class is dominated by at least one vertex. The dominated chromatic number of G, denoted by $\chi_{dom}(G)$, is the minimum number of colors among all dominated colorings of G.*

More results on the dominated coloring can be found in [32–34].

For problems mentioned above, the domination property is defined either on vertices or on color classes. Indeed, each color class in a dominator coloring is not necessarily dominated by a vertex, and each vertex in a dominated coloring does not necessarily dominate a color class. In this paper, we introduce the domination coloring that both of the vertices and color classes should satisfy the domination property.

Definition 3. *A domination coloring of a graph G is a proper vertex coloring of G, such that each vertex of G dominates at least one color class (possibly its own class), and each color class is dominated by at least one vertex. The domination chromatic number of G, denoted by $\chi_{dd}(G)$, is the minimum number of color classes in a domination coloring of G.*

The domination coloring problem is to find a domination coloring of G, such that the number of color classes is minimized. Here, we describe a possible application for the domination coloring problem in the following scenario. In a social network, social actors are represented as vertices and their relationships as edges (two actors are adjacent if they are friends). Two strangers can become friends by their mutual friend (i.e., intermediary). Then, each actor wants to develop interpersonal relationships in the social network by some intermediaries, meanwhile, each actor wants to be the important intermediary of other strangers. The domination coloring problem involves finding the minimum groups of actors in the social network with the below properties:

1. Actors in the same group are strangers;
2. Actors in the same group can become friends by at least one common intermediary;
3. Each actor is an intermediary of at least one actor (stranger) group.

We proceed as follows. In the rest of Section 1, we recall some basic definitions that will be used in the following sections. In Section 2, we analyse the complexity of the k-domination coloring problem. In Section 3, we present basic results and properties of the domination chromatic number $\chi_{dd}(G)$, including the bounds and characterization results. In Section 4, we further research $\chi_{dd}(G)$ of some special classes of graphs, including the split graphs, the generalized Petersen graphs $P(n,1)$, corona products, and edge corona products. In Section 5, we investigate some realization results on graphs with $\chi_{dd}(G) = \chi(G)$. Finally, we make a conclusion in Section 6.

1.2. Preliminary

Graphs considered in this paper are finite, simple, undirected, and connected. Let $G = (V, E)$ be a graph with $n = |V|$ and $m = |E|$. For any vertex $v \in V(G)$, the open neighborhood of v is the set $N(v) = \{u|uv \in E(G)\}$ and the closed neighborhood is the set $N[v] = N(v) \cup \{v\}$. Similarly, the open and closed neighborhoods of a set $X \subseteq V$ are, respectively, $N(X) = \bigcup_{x \in X} N(x)$ and $N[X] = N(X) \cup X$. The degree of a vertex $v \in V$, denoted by $deg(v)$, is the cardinality of its open neighborhood. The maximum and minimum degree of a graph G is denoted by $\Delta(G)$ and $\delta(G)$, respectively. We call a vertex of degree one a leaf or a pendant vertex, its adjacent vertex a support vertex. Given a set $X \subseteq V$, we denote by $G[X]$ the subgraph of G induced by X. Given any graph H, a graph G is H-free if it does not have any induced subgraph isomorphic to H. We denote by P_n the path on n vertices and by C_n the cycle on n vertices. A tree is a connected acyclic graph. The complete graph on n vertices is denoted by K_n and the complete graph of order 3 is called a triangle. The complete bipartite graph with classes of orders r and s is denoted by $K_{r,s}$. A star S_k is the graph $K_{1,k}$ with $k \geq 1$.

An independent set in G is a set of vertices, such that any two vertices in the set are not adjacent. A matching in a graph G is a set of nonadjacent edges of G. The matching number $\alpha'(G)$ is the cardinality of a largest matching in G. A vertex cover in a graph G is a set of vertices, such that each edge has at least one endpoint in the set. The vertex cover number $\beta(G)$ is the cardinality of a smallest vertex cover in G. The clique number $w(G)$ of a graph G is the maximum order among the complete subgraphs of G.

A proper vertex k-coloring of a graph $G = (V, E)$ is a mapping $f : V \to \{1, 2, \cdots, k\}$, such that any two adjacent vertices receive different colors. In fact, this problem is equivalent to the problem of partitioning the vertex set of G into k independent sets $\{V_1, V_2, \cdots, V_k\}$ where $V_i = \{x \in V | f(x) = i\}$. The set of all vertices colored with the same color is called a color class. The chromatic number of G, denoted by $\chi(G)$, is the minimum number of colors among all proper colorings of G.

A dominating set S is a subset of the vertices in a graph G, such that every vertex in G either belongs to S or has a neighbor in S. The domination number $\gamma(G)$ is the minimum cardinality of a dominating set of G. A $\gamma(G)$-set is a dominating set of G with minimum cardinality.

For any undefined terms, the reader is referred to the book by Bondy and Murty [35].

2. Complexity Results

This section focuses on the complexity study of the domination coloring problem, e.g., whether an arbitrary graph admits a domination coloring with the most k colors. We give the formalization of this problem.

- k-domination coloring problem.
  Instance: a graph $G = (V, E)$ without isolated vertices and a positive integer k.
  Question: is there a domination coloring of G with the most k colors?

Theorem 1. *For $k \geq 4$, the k-domination coloring problem is NP-complete.*

Proof. The k-domination coloring problem is in NP, since verifying if a coloring is a domination coloring could be performed in polynomial time. Now, we give a polynomial time reduction from the k-coloring problem, which is known to be NP-complete, for $k \geq 3$. Let $G = (V, E)$ be a graph without isolated vertices. We construct a graph G' from G by adding a new vertex x to G and adding edges between x and every vertex of G. That is, x is a dominating vertex of G', as shown in Figure 1. We show that G admits a proper coloring with k colors if and only if G' admits a domination coloring with $k + 1$ colors.

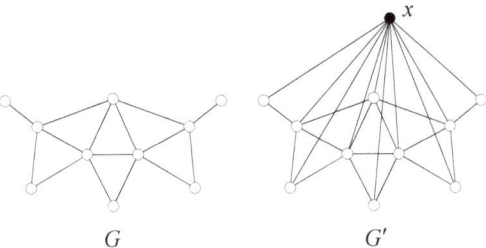

Figure 1. The graphs G and G'.

First, we prove the necessity. Let f be a proper k-coloring of G, and the corresponding color classes set is $\{V_1, V_2, \cdots, V_k\}$. We construct a $(k+1)$-domination coloring f' of G' with the color classes set $\{V'_1 = V_1, V'_2 = V_2, \cdots, V'_k = V_k, V'_{k+1} = \{x\}\}$. It is easy to see that f' is a domination coloring of G' since

1. f' is proper;
2. Each vertex other than x dominates at least the color class containing x and x dominates all color classes of f';
3. Each color class other than $\{x\}$ is dominated by x and the color class containing x is dominated by any other vertex.

Then, we prove the sufficiency. Let f' be a $(k+1)$-domination coloring of G', and $\{V'_1, V'_2, \cdots, V'_k, V'_{k+1}\}$ is the color classes set. Since f' is proper, there exists a color class V'_i

such that $V'_i = \{x\}$. Thus, we can construct a proper k-coloring of G by removing the color class V'_i from f'.

From the above, the k-domination coloring problem is NP-complete, for $k \geq 4$. □

3. Basic Results and Properties of the Domination Chromatic Number

In this section, we study some properties of the domination coloring and basic results on typical classes of graphs.

Let G be a connected graph with order $n \geq 2$. Then at least two different colors are needed in a domination coloring since there are at least two vertices in G adjacent to each other. Moreover, if each vertex receives a unique color, then both the vertices and color classes satisfy the domination property. Clearly, we get a domination coloring of G with n colors. Thus,

$$2 \leq \chi_{dd}(G) \leq n. \tag{1}$$

Gera et al. [20] introduced the Inequalities (2) for the dominator chromatic number $\chi_d(G)$ and Merouane et al. [31] obtained Inequalities (3) for the dominated chromatic number $\chi_{dom}(G)$. Moreover, we can get a similar inequality for the domination chromatic number $\chi_{dd}(G)$.

$$\max\{\chi(G), \gamma(G)\} \leq \chi_d(G) \leq \chi(G) + \gamma(G). \tag{2}$$

$$\max\{\chi(G), \gamma(G)\} \leq \chi_{dom}(G) \leq \chi(G) \cdot \gamma(G). \tag{3}$$

Proposition 1. *Let G be a graph without isolated vertices, then*

$$\max\{\chi(G), \gamma(G)\} \leq \max\{\chi_d(G), \chi_{dom}(G)\} \leq \chi_{dd}(G) \leq \chi(G) \cdot \gamma(G).$$

Proof. Since any domination coloring of G is also a dominator coloring and a dominated coloring, $\max\{\chi_d(G), \chi_{dom}(G)\} \leq \chi_{dd}(G)$. Both the dominator coloring and dominated coloring are proper vertex colorings of G, so, $\chi(G) \leq \max\{\chi_d(G), \chi_{dom}(G)\}$. For any dominator coloring (dominated coloring) of G, we can get a dominating set by taking a vertex in each color class. Thus, $\gamma(G) \leq \max\{\chi_d(G), \chi_{dom}(G)\}$. Therefore, the left two parts of the inequality hold.

For the right part of the inequality, we consider a $\gamma(G)$-set D of G. A domination coloring of G can be obtained by giving distinct colors to each vertex x of D and at most $\chi(G) - 1$ new colors to the vertices of $N(x)$. Hence, we totally use at most $\gamma(G) + (\chi(G) - 1) \cdot \gamma(G) = \chi(G) \cdot \gamma(G)$ colors. So, $\chi_{dd}(G) \leq \chi(G) \cdot \gamma(G)$. □

The bound of Proposition 1 is tight for complete graphs. Since every planar graph is "4-colorable" [2,3], the following result is straightforward:

Corollary 1. *Let G be a planar graph without isolated vertices, then $\chi_{dd}(G) \leq 4\gamma(G)$.*

Proposition 2. *Let G be a connected graph with order n and maximum degree Δ, then $\chi_{dd}(G) \geq \frac{n}{\Delta}$.*

Proof. Consider a minimum domination coloring of G. Since G is $S_{\Delta+1}$-free, any color class would not have more than Δ vertices; otherwise, a vertex dominating such a color class will induce a star of order at least $\Delta + 2$, a contradiction. So, $\chi_{dd}(G) \geq \frac{n}{\Delta}$. □

Theorem 2. *Let G be a connected triangle-free graph, then $\chi_{dd}(G) \leq 2\gamma(G)$.*

Proof. Consider a minimum dominating set S of G. Color every vertex of S with a new color. Since G does not contain any triangle, the set of neighbors of every vertex of S is an independent set. Thus, a second new color is given for each neighborhood. Obviously, this is a proper coloring of G with $2|S|$ colors, which satisfies that every vertex dominates at least one color class, and every color class is dominated by at least one vertex. Thus, $\chi_{dd}(G) \leq 2\gamma(G)$. □

Theorem 3. *(1) For the path P_n, $n \geq 2$,*

$$\chi_{dd}(P_n) = 2 \cdot \lfloor \tfrac{n}{3} \rfloor + \mathrm{mod}(n,3);$$

(2) For the cycle C_n,

$$\chi_{dd}(C_n) = \begin{cases} 2, & n = 4, \\ 3, & n = 3, 5, \\ 2 \cdot \lfloor \tfrac{n}{3} \rfloor + \mathrm{mod}(n,3), & \text{otherwise}; \end{cases}$$

(3) For the complete graph K_n, $\chi_{dd}(K_n) = n$;
(4) For the complete k-partite graph $K_{a_1,a_2,\cdots,a_k}$, $\chi_{dd}(K_{a_1,a_2,\cdots,a_k}) = k$;
(5) For the complete bipartite graph $K_{r,s}$, $\chi_{dd}(K_{r,s}) = 2$;
(6) For the star $K_{1,n}$, $\chi_{dd}(K_{1,n}) = 2$;
(7) For the wheel $W_{1,n}$,

$$\chi_{dd}(W_{1,n}) = \begin{cases} 3, & n \text{ is even}, \\ 4, & n \text{ is odd}. \end{cases}$$

Proof. (1) Let $P_n = v_1 v_2 \cdots v_n$. By the definition of the domination coloring, we discover that at most two non-adjacent vertices are allowed in a color class, if not, there exist no vertex dominating this color class. On the other hand, the vertex adjacent to both vertices of a color class must be the unique vertex of some color class. For convenience, let $P_5 = v_1 v_2 v_3 v_4 v_5$ be a P_5-subgraph of P_n. If vertices v_1 and v_3 are in a color class, then v_2 must be the unique vertex of a color class. If not, v_2 and v_4 are partitioned in a color class, which will result in v_4 cannot dominate any color class. Thus, every three vertices of P_n need to be partitioned in two color classes, and the rest form their own color class. Clearly, it is an optimal domination coloring of P_n. Thus, $\chi_{dd}(P_n) = 2 \cdot \lfloor \tfrac{n}{3} \rfloor + \mathrm{mod}(n,3)$.

(2) For $n = 3, 4, 5$, the result follows by inspection. For $n \geq 6$, it is not hard to find the case is similar to the path P_n. As the discussion in (1), the result follows.

(3) For the complete graph K_n, $\chi(K_n) = n$. By Proposition 1 and in Equation (1), $\chi_{dd}(K_n) = n$.

(4) Let $K_{a_1,a_2,\cdots,a_k}$ be the complete k-partite graph, and $V_i (1 \leq i \leq k)$ be the k-partite sets. Then $\chi_{dd}(K_{a_1,a_2,\cdots,a_k}) \geq \chi(K_{a_1,a_2,\cdots,a_k}) = k$. Moreover, the coloring that assigns color i to each partite set $V_i (1 \leq i \leq k)$ is a domination coloring. The result follows.

(5) and (6) are special cases of (4).

(7) Let $W_{1,n}$ be the wheel with order $n+1$. Since,

$$\chi(W_{1,n}) = \begin{cases} 3, & n \text{ is even}, \\ 4, & n \text{ is odd}. \end{cases}$$

and the corresponding proper colorings are also domination colorings, the result follows. □

Note. For a given graph G, and a subgraph H of G, the domination chromatic number of H can be smaller or larger than the domination chromatic number of G. That is to say, induction may be not useful when we want to find the domination chromatic number of a graph. As an example, consider the graph $G = K_n$ and $H = P_2$, then $\chi_{dd}(K_n) = n \geq 2 = \chi_{dd}(P_2)$, and consider the graph $G = K_{n,n}$ and $H = P_{2n}$, then $\chi_{dd}(K_{n,n}) = 2 \leq 2 \cdot \lfloor \tfrac{2n}{3} \rfloor + \mathrm{mod}(2n,3) = \chi_{dd}(P_{2n})$.

Theorem 4. *For the Petersen graph P, $\chi_{dd}(P) = 5$.*

Proof. It is easy to check $\{\{v_1, v_2, v_9\}, \{v_3, v_4, v_6\}, \{v_5, v_7\}, \{v_8\}, \{v_{10}\}\}$ is a domination coloring of the Petersen graph, as shown in Figure 2. So, $\chi_{dd}(P) \leq 5$. By Proposition 2.1

in [32], $\chi_{dom}(P) = 4$ and $\chi_d(P) = 5$. Then, $\chi_{dd}(P) \geq 5$ by Proposition 1. Therefore, $\chi_{dd}(P) = 5$. □

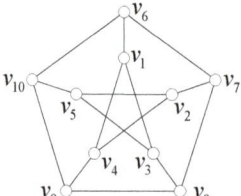

Figure 2. The Petersen graph.

Next, we consider the bi-stars. Let $S_{p,q}$ be the bi-star with central vertices u and v, where $deg(u) = p \geq 2$ and $deg(v) = q \geq 2$. Let $X = \{x_1, x_2, \cdots, x_{p-1}\}$ and $Y = \{y_1, y_2, \cdots, y_{q-1}\}$. Obviously, $N(u) = X \cup \{v\}$ and $N(v) = Y \cup \{u\}$, as shown in Figure 3.

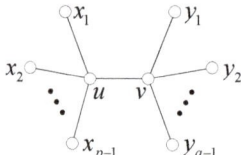

Figure 3. The bi-star $S_{p,q}$.

Theorem 5. *For the bi-star $S_{p,q}$ with $p + q \geq 5$, $\chi_{dd}(S_{p,q}) = 4$.*

Proof. Consider a proper coloring of $S_{p,q}$ in which the color classes $V_1 = \{u\}$, $V_2 = \{v\}$, $V_3 = X$, and $V_4 = Y$. Then, each vertex in the set $\{u\} \cup X$ dominates the color class V_1, and each vertex in the set $\{v\} \cup Y$ dominates the color class V_2. Moreover, the color class V_1 is dominated by any vertex in V_3, V_2 is dominated by any vertex in V_4, V_3 is dominated by vertex u, and V_4 is dominated by vertex v. Therefore, this is a domination coloring, and $\chi_{dd}(S_{p,q}) \leq 4$.

By the Lemma 2.2 in [20], $\chi_d(S_{p,q}) = 3$. So, $3 \leq \chi_{dd}(S_{p,q}) \leq 4$. Suppose that $\chi_{dd}(S_{p,q}) = 3$. It will be result in that each vertex in X or each vertex in y does not dominate a color class. Thus, $\chi_{dd}(S_{p,q}) = 4$. □

Theorem 6. *Let G be a connected graph with order n. Then $\chi_{dd}(G) = 2$ if and only if $G = K_{r,s}$ for $r, s \in \mathbf{N}$.*

Proof. By Theorem 3 (5), if $G = K_{r,s}$, then $\chi_{dd}(G) = 2$. We just need to prove the necessity.

Let G be a connected graph, such that $\chi_{dd} = 2$, and V_1 and V_2 are the two color classes. If $|V_1| = 1$ or $|V_2| = 1$, then $G = K_{1,n-1}$. So, suppose that $|V_1| \geq 2$ and $|V_2| \geq 2$. For any vertex $x \in V_1$, since $|V_1| \geq 2$, it follows that x dominates color class V_2. Similarly for any vertex in V_2. Thus, each vertex of V_1 is adjacent to each vertex of V_2, and both V_1 and V_2 are independent. So $G = K_{r,s}$ for $r, s \in \mathbf{N}$, and the result follows. □

Theorem 7. *Let G be a connected graph with order n. Then $\chi_{dd}(G) = n$ if and only if $G = K_n$ for $n \in \mathbf{N}$.*

Proof. By Theorem 3 (3), $\chi_{dd}(G) = n$, if $G = K_n$. We only need to prove the necessity.

Let G be a connected graph with $\chi_{dd}(G) = n$. Suppose that $G \neq K_n$. Thus, there exist two vertices, say x and y, such that they are not adjacent and they have a common neighbor in G. Now, we define a coloring of G in which x and y receive the same color, and each of the remaining vertices receive a unique color. This is a domination coloring, so $\chi_{dd}(G) \leq n - 1$, a contradiction. Thus, $G = K_n$, and we obtain the result. □

4. Domination Coloring in Some Classes of Graphs

In this section, we further research the domination coloring of some classes of graphs, including the split graphs, the generalized Petersen graphs $P(n,1)$, corona products, and edge corona products.

4.1. Domination Coloring for Split Graphs

We study the domination chromatic number of split graphs in this subsection.

A graph G is called a split graph if its vertex set can be partitioned into a clique and an independent set.

Theorem 8. *Let G be a split graph with split partition (K, I) and its maximum clique is of order k. If there exists a dominating set D of G, such that $D \subseteq K$, and every vertex in I is adjacent to at least one vertex in $K - D$ and nonadjacent to at least one vertex in $K - D$, then $\chi_{dd}(G) = k$.*

Proof. Consider a minimum domination coloring of G. Obviously, $\chi_{dd}(G) \geq k$. We give now a construction that yields a domination coloring of G with k colors.

First, we give to each vertex of D a unique new color from the set $\{1, \ldots, p\}$ and each vertex of $K - D$ a unique new color from the set $\{1, \ldots, q\}$, where $p + q = k$. We arrange the vertices in $K - D$ according to a circular order function defined on the set $\{1, \ldots, q\}$ as follows:
$$j \in \{1, \ldots, q\} \implies next(j) = j \bmod(k) + 1.$$

We now color the vertices of the independent set I of the split graph G. Let i be a vertex of I and let $N(i)$ be the set (of colors) of its neighbors. The color of i is given by the following formula:
$$c(i) = \min\{j : j \in \{\{1, \ldots, q\} \setminus N(i)\} \wedge next(j) \in N(i)\}.$$

Since every vertex i in I is adjacent to at least one vertex in $K - D$ and nonadjacent to at least one vertex in $K - D$, at least one color from the set $\{1, \ldots, q\}$ would be available for i. Thus, every vertex in G is properly colored. On the one hand, given that $D \subseteq K$ is a dominating set of G, each vertex dominates a color class formed by a vertex of D. On the other hand, from the above construction, each color class formed by a vertex of D is obviously dominated, and one can observe that each color j will appear only in the neighborhood of the vertex from the clique colored with the color $next(j)$. Thus, we obtain that the proposed construction gives a domination coloring for the split graph G with k colors. □

4.2. Domination Coloring for Generalized Petersen Graphs $P(n,1)$

In this subsection, we determine the domination chromatic number of the generalized Petersen graph $P(n,1)$.

Let n and k be positive integers with $n \geq 3$ and $k \leq n - 1$. The generalized Petersen graph $P(n,k)$ is the graph with $V(P(n,k)) = \{u_1, u_2, u_3, \cdots, u_n\} \cup \{v_1, v_2, v_3, \cdots, v_n\}$ and $E(P(n,k)) = \{v_i v_{i+1} : 1 \leq i \leq n\} \cup \{v_i u_i : 1 \leq i \leq n\} \cup \{u_i u_{i+k} : 1 \leq i \leq n\}$ where the addition in the subscript is modulo n.

The Cartesian product $G \square H$ of two graphs G and H is the graph with $V(G \square H) = V(G) \times V(H)$ and $E(G \square H) = \{(g_1, h_1)(g_2, h_2) : g_1 = g_2 \text{ and } h_1 h_2 \in E(H) \text{ or } h_1 = h_2 \text{ and } g_1 g_2 \in E(G)\}$.

The generalized Petersen graph $P(n,1)$ is isomorphic to the Cartesian product $C_n \square K_2$. We now proceed to determine $\chi_{dd}(P(n,1))$.

Theorem 9. *For the generalized Petersen graph $P(n,1)$, we have*

$$\chi_{dd}(P(n,1)) = \begin{cases} n, & n \equiv 0 \pmod 4, \\ n+1, & \text{otherwise}. \end{cases}$$

Proof. The result is obvious when $n = 3$. Now, let $n = 4k + j$ where $k \geq 1$ and $0 \leq j \leq 3$. Let $\mathcal{S} = \{N[v_{4i-3}] : 1 \leq i \leq k\} \cup \{N[u_{4i-1}] : 1 \leq i \leq k\}$. Then $\mathcal{S}$ is a family of $2k$ disjoint closed neighborhoods in $P(n,1)$. We consider the following cases:

Case 1. $j = 0$.

In this case, $\mathcal{S}$ covers all the vertices of $P(n,1)$. For each closed neighborhood $N[x]$ in $\mathcal{S}$, give a color to the vertex x and another color to the neighbors of x. Obviously, $\mathcal{C} = \{\{v_{4i-3}\} : 1 \leq i \leq k\} \cup \{\{N(v_{4i-3})\} : 1 \leq i \leq k\} \cup \{\{u_{4i-1}\} : 1 \leq i \leq k\} \cup \{\{N(u_{4i-1})\} : 1 \leq i \leq k\}$ is a domination coloring of $P(n,1)$. Thus, $\chi_{dd}(P(n,1)) \leq 4k = n$. On the other hand, any two disjoint closed neighborhoods in $\mathcal{S}$ cannot have a common color, which will result in some vertices having no color class to dominate, and some color classes will not dominate by any vertex. In this sense, $\chi_{dd}(P(n,1)) \geq 4k = n$. Therefore, $\chi_{dd}(P(n,1)) = 4k = n$.

Case 2. $j = 1$.

In this case, $\mathcal{S}$ is a collection of $2k$ disjoint closed neighborhoods in $P(n,1)$ and the vertices v_{n-1} and u_n are not covered by $\mathcal{S}$. Similar to Case 1, give two colors to each closed neighborhood in $\mathcal{S}$ and two new colors to v_{n-1} and u_n. Then, $\mathcal{C} = \{\{v_{4i-3}\} : 1 \leq i \leq k\} \cup \{\{N(v_{4i-3})\} : 1 \leq i \leq k\} \cup \{\{u_{4i-1}\} : 1 \leq i \leq k\} \cup \{\{N(u_{4i-1})\} : 1 \leq i \leq k\} \cup \{v_{n-1}\} \cup \{u_n\}$ is a domination coloring of $P(n,1)$. Thus, $\chi_{dd}(P(n,1)) \leq 4k+2 = n+1$. On the other hand, to ensure that every vertex dominate a color class and every color class is dominated by a vertex, the vertices v_{n-1} and u_n should be colored uniquely, respectively. Hence, $\chi_{dd}(P(n,1)) \geq 4k+2 = n+1$. Therefore, $\chi_{dd}(P(n,1)) = 4k+2 = n+1$.

Case 3. $j = 2$.

In this case, $\mathcal{S}$ is a collection of $2k$ disjoint closed neighborhoods in $P(n,1)$ and the vertices $v_{n-2}, v_{n-1}, u_{n-1}$ and u_n are not covered by $\mathcal{S}$. $\mathcal{C} = \{\{v_{4i-3}\} : 1 \leq i \leq k\} \cup \{\{N(v_{4i-3})\} : 1 \leq i \leq k\} \cup \{\{u_{4i-1}\} : 1 \leq i \leq k\} \cup \{\{N(u_{4i-1})\} : 1 \leq i \leq k\} \cup \{v_{n-2}, u_{n-1}\} \cup \{v_{n-1}\} \cup \{u_n\}$ is a domination coloring of $P(n,1)$. Thus, $\chi_{dd}(P(n,1)) \leq 4k+3 = n+1$. On the other hand, to ensure the domination properties, the vertices $v_{n-2}, v_{n-1}, u_{n-1}$ and u_n need at least three new colors. Hence, $\chi_{dd}(P(n,1)) \geq 4k+3 = n+1$. Therefore, $\chi_{dd}(P(n,1)) = 4k+3 = n+1$.

Case 4. $j = 3$.

In this case, $\mathcal{S} \cup N[v_{4k+1}]$ is a collection of $2k+1$ disjoint closed neighborhoods in $P(n,1)$ and the vertices u_{n-1} and u_n are not covered by these neighborhoods. Then $\mathcal{C} = \{\{v_{4i-3}\} : 1 \leq i \leq k\} \cup \{\{N(v_{4i-3})\} : 1 \leq i \leq k\} \cup \{\{u_{4i-1}\} : 1 \leq i \leq k\} \cup \{\{N(u_{4i-1})\} : 1 \leq i \leq k\} \cup \{v_{4k+1}\} \cup \{N(v_{4k+1})\} \cup \{u_{n-1}\} \cup \{u_n\}$ is a domination coloring of $P(n,1)$. Thus, $\chi_{dd}(P(n,1)) \leq 4k+4 = n+1$. Similar to the above analysis, $\chi_{dd}(P(n,1)) \geq 4k+4 = n+1$. Therefore, $\chi_{dd}(P(n,1)) = 4k+4 = n+1$.

Thus, the result follows. □

4.3. Domination Coloring for Corona Products

For graphs G and H, the corona product $G \circ H$ is obtained from one copy of G and $n(G)$ copies of H by joining with an edge each vertex of the ith copy of H, $i \in [n(G)]$, to the ith vertex of G. If $v \in V(G)$, then the copy of H in $G \circ H$ corresponding to v will be denoted by H_v. We may consider the vertex set of $G \circ H$ to be

$$V(G \circ H) = V(G) \cup (\bigcup_{v \in V(G)} V(H_v)).$$

The dominator and dominated chromatic numbers of corona products are already known.

Theorem 10 ([36]). *If G and H are graphs, then $\chi_d(G \circ H) = n(G) + \chi(H)$.*

Theorem 11 ([33]). *If G and H are graphs, then $\chi_{dom}(G \circ H) = n(G)\chi(H)$.*

We now give a general result for the domination chromatic number of corona products.

Theorem 12. *If G and H are graphs, then $\chi_{dd}(G \circ H) \leq n(G)(\chi(H) + 1)$.*

Proof. Set $n = n(G)$ and color $G \circ H$ as follows. First, we color each vertex of $V(G)$ an unique color. Second, we properly color every copy of H with $\chi(H)$ distinct colors. Clearly, the obtained coloring is a domination coloring of $G \circ H$. Indeed, each vertex $v \in V(G)$ forms a color class of cardinality 1 and the color class $\{v\}$ is dominated by any adjacent vertices of v, while each vertex from H_v is adjacent to the vertex v and dominate the color class $\{v\}$, the color class formed by vertices in H_v is dominated by the corresponding vertex v. Therefore, $\chi_{dd}(G \circ H) \leq n(G)(\chi(H) + 1)$. □

4.4. Domination Coloring for Edge Corona Products

For graphs G and H, the edge corona $G \diamond H$ is obtained by taking one copy of G and $m(G)$ disjoint copies of H one-to-one assigned to the edges of G, and for every edge $vv' \in E(G)$ joining v and v' to every vertex of the copy of H associated to vv'. If $e = vv' \in E(G)$, then the copy of H in $G \diamond H$ corresponding to vv' will be denoted with $H_{vv'}$ (or simply H_e). Hence we may consider the vertex set of $G \diamond H$ to be

$$V(G \diamond H) = V(G) \cup \left(\bigcup_{vv' \in E(G)} V(H_{vv'}) \right).$$

The dominator and dominated chromatic numbers of edge corona products have been studied, which were related to the matching number α' and the vertex cover number β.

Theorem 13 ([36]). *If G and H are graphs, then $\chi_d(G \diamond H) = \beta(G) + \chi(H) + 1$.*

Theorem 14 ([36]). *If G is a graph without pendant vertices, then $\chi_{dom}(G \diamond H) \geq \alpha'(G)\chi(H) + \chi_{dom}(G)$.*

Theorem 15 ([36]). *If G has k pendant vertices, then $\chi_{dom}(G \diamond H) \geq \alpha'(G)\chi(H) + k$.*

Theorem 16 ([36]). *If G and H are graphs, then $\chi_{dom}(G \diamond H) \leq \beta(G)\chi(H) + \chi_{dom}(G)$, with equality when G is bipartite graph without pendant vertices.*

In the following, we give a general result for the domination chromatic number of edge corona products.

Theorem 17. *If G and H are graphs, then $\chi_{dd}(G \diamond H) \leq \beta(G)(\chi(H) + 2)$.*

Proof. Let $K = \{v_1, \cdots, v_{\beta(G)}\}$ be a minimum vertex cover of G, so that $|K| = \beta(G)$. Partition $E(G)$ into subsets of edges $E_1, \cdots, E_{\beta(G)}$, such that if $e \in E_i$, then v_i is an endpoint of e, $i = 1, \cdots, \beta(G)$. Partition $V - K$ into subsets of vertices $V_1, \cdots, V_{\beta(G)}$, such that if $u \in V_i$, then $uv_i \in E_i$, $i = 1, \cdots, \beta(G)$. It is clear that such partitions always exists since K is a vertex cover. Notice that each V_i is an independent set, $i = 1, \cdots, \beta(G)$.

Now, define a coloring c of $G \diamond H$ as follows. First, for each set of edges E_i, reserve private $\chi(H)$ colors and color with each of the corresponding subgraphs H_e, $e \in E_i$. Second, color the vertices of K with $\beta(G)$ colors. Third, color the vertices of $V_1, \cdots, V_{\beta(G)}$ with additional $\beta(G)$ colors. Then, c is the domination coloring of $G \diamond H$. Indeed, each vertex in each H_e dominate a corresponding color class $\{v_i\}$, and the color classes of those copies of H with common $\chi(H)$ colors are dominated by the corresponding vertex v_i. For each vertex v_i in K, there exist color classes dominated by v_i. Moreover, the color class $\{v_i\}$ can be dominated by any adjacent vertex of v_i. Moreover, each vertex in V_i

dominates the color class $\{v_i\}$, and the color class V_i is dominated by vertex v_j. Hence, $\chi_{dd}(G \diamond H) \leq \beta(G)(\chi(H) + 2)$. □

5. Graphs with $\chi_{dd}(G) = \chi(G)$

For any graph G, we have $\chi_{dd}(G) \geq \chi(G)$. In this section, we investigate graphs for which $\chi_{dd}(G) = \chi(G)$.

The following theorem directly follows from Proposition 1.

Theorem 18. *Let G be a connected graph, if $\gamma(G) = 1$, then $\chi_{dd}(G) = \chi(G)$.*

A unicyclic graph is a graph that contains only one cycle. In the following, we characterize unicyclic graphs with $\chi_{dd} = \chi$.

Theorem 19. *Let G be a connected unicyclic graph. Then $\chi_{dd}(G) = \chi(G)$ if and only if G is isomorphic to C_3 or C_4 or C_5 or the graph obtained from C_3 by attaching any number of leaves at one vertex of C_3.*

Proof. For the sufficiency, the result is obvious if G is the graph meet conditions. We consider only the necessity. Let G be a connected unicyclic graph with $\chi_{dd}(G) = \chi(G)$, and C the unique cycle of G.

Case 1. If C is an even cycle, then $\chi(G) = 2$ and $\chi_{dd}(G) = 2$. It follows that G cannot contain any other vertices not on C, otherwise $\chi_{dd}(G) \geq 3$. By Theorem 3(2), $G = C_4$.

Case 2. If C is an odd cycle, then $\chi_{dd}(G) = \chi(G) = 3$. Suppose there exists a support vertex x not on C. Since x or the leaf is a color class in each χ_{dd}-coloring of G, it follows that $\chi_{dd}(G) \geq 4$, which is a contradiction. Hence, all of the support vertices lie on C, and any vertex not on C is a leaf. Moreover, the number of support vertices is at most one. Otherwise, it follows that some color classes are not dominated, since there exists some χ_{dd}-coloring of G in which every support vertex appears as a singleton color class.

Case 2.1. If $|C| = 3$, then G is isomorphic to C_3 or the graph obtained from C_3 by attaching any number of leaves at exactly one vertex of C_3.

Case 2.2. Suppose that $|C| \geq 5$. If there exists a support vertex x on C, then there exists a χ_{dd}-coloring $\{V_1, V_2, \{x\}\}$ of G, such that V_1 contains all of the leaves of x. Now, we get two vertices u and v on C, such that $u \in V_1, v \in V_2$, both u and v are not adjacent to x. Clearly, v does not dominate any color class and the color class V_1 is not dominated by any vertex, which is a contradiction. Thus, G has no support vertices and $G = C$. By Theorem 3(2), $G = C_5$. So, the theorem follows. □

For the complete graph K_n, we know that $\chi_{dd}(K_n) = \chi(K_n) = n$. Next, we construct a family of graphs by attaching leaves at some vertices of the complete graph. We denote by $\mathcal{K}_n^m$ the family of graphs obtained by attaching leaves at m vertices of K_n, $1 \leq m \leq n$. We take no account of the number of leaves attached at any vertex in the notation, since it does not impact the domination chromatic number. Moreover, we denote any element in $\mathcal{K}_n^m$ by K_n^m. For example, a instance of K_5^2 is shown in Figure 4.

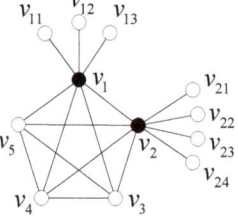

Figure 4. A instance of K_5^2.

Theorem 20. *For* $m \leq \lfloor \frac{n}{2} \rfloor$, $\chi_{dd}(K_n^m) = \chi(K_n^m)$.

Proof. For any $1 \leq m \leq n$, K_n^m is n-colorable. So, $\chi(K_n^m) = n$. Next, we consider a domination coloring of K_n^m. On the one hand, vertex-attached leaves should be partitioned into a singleton color class, since each leaf has to dominate a color class formed by its only neighbor. On the other hand, leaves attached to different vertices have to be partitioned into different color classes, otherwise, there exists no vertex dominating the color class. Thus, at most $\lfloor \frac{n}{2} \rfloor$ vertices can be attached to leaves of K_n, in order to guarantee that K_n^m is n-domination colorable. The result follows. □

6. Conclusions

In this paper, we introduce the concept of domination coloring where both vertices and color classes should satisfy the domination property. Moreover, an application of domination coloring in a social network scenario is presented. We prove the k-domination coloring problem is NP-complete by a reduction from the k-coloring problem. We provide basic results and properties of the domination chromatic number $\chi_{dd}(G)$, and further research $\chi_{dd}(G)$ of the split graphs, the generalized Petersen graphs $P(n,1)$, the corona products, and edge corona products. In particular, we establish a relationship between the domination chromatic number and other graph parameters, such as the matching number, the vertex cover number, and the clique number. Moreover, we provide sufficient and necessary conditions for connected unicyclic graphs with $\chi_{dd} = \chi$, and construct a class of graphs with $\chi_{dd} = \chi$. Our future work will focus on the relationships among the domination chromatic number, the domination number, and the chromatic number, and discuss graphs with $\chi_{dd} = \chi$, $\chi_{dd} = \gamma$, $\chi_{dd} = \chi \cdot \gamma$. Moreover, we will explore the application of domination coloring in practice.

Author Contributions: Created and conceptualized the idea, Y.Z. and J.X.; writing—original draft preparation, Y.Z. and D.Z.; writing—review and editing, Y.Z. and M.M. All authors have read and agreed to the published version of the manuscript.

Funding: This research was supported by the National Key R&D Program of China no. 2019YFA0706401; the National Natural Science Foundation of China General program no. 62172014, no. 62172015, no. 61872166; and the National Natural Science Foundation of China Youth Program no. 62002002.

Institutional Review Board Statement: Not applicable.

Informed Consent Statement: Not applicable.

Data Availability Statement: Not applicable.

Conflicts of Interest: The authors declare no conflict of interest.

References

1. Franklin, P. The four color problem. *Am. J. Math.* **1922**, *44*, 225–236. [CrossRef]
2. Appel, K.; Haken, W. Every planar map is four colorable. part i: Discharging. *Ill. J. Math.* **1977**, *21*, 429–490. [CrossRef]
3. Appel, K.; Haken, W.; Koch, J. Every planar map is four colorable. part ii: Reducibility. *Ill. J. Math.* **1977**, *21*, 491–567. [CrossRef]
4. Pardalos, P.M.; Mavridou, T.; Xue, J. *The Graph Coloring Problem: A Bibliographic Survey*; Springer: Boston, MA, USA, 1998.
5. Malaguti, E.; Toth, P. A survey on vertex coloring problems. *Int. Trans. Oper. Res.* **2010**, *17*, 1–34. [CrossRef]
6. Borodin, O.V. Colorings of plane graphs: A survey. *Discret. Math.* **2013**, *313*, 517–539. [CrossRef]
7. Li, G.Z.; Simha, R. The partition coloring problem and its application to wavelength routing and assignment. In Proceedings of the First Workshop on Optical Networks, Dallas, TX, USA, 1 May 2000; pp. 1–19.
8. Zhu, E.Q.; Jiang, F.; Liu C.J.; Xu, J. Partition independent set and reduction-based approach for partition coloring problem. *IEEE Trans. Cybern.* **2020**, 1–10. [CrossRef] [PubMed]
9. MacGillivray, G.; Seyffarth, K. Domination numbers of planar graphs. *J. Graph Theory* **1996**, *22*, 213–229. [CrossRef]
10. Haynes, T.W.; Hedetniemi, S.T.; Slater, P.J. *Fundamentals of Domination in Graphs*; Marcel Dekker, Inc.: New York, NY, USA, 1998.
11. Haynes, T.W.; Hedetniemi, S.T.; Slater, P.J. *Domination in Graphs: Volume 2: Advanced Topics*; Marcel Dekker, Inc.: New York, NY, USA, 1998.
12. Honjo, T.; Kawarabayashi, K.-I.; Nakamoto, A. Dominating sets in triangulations on surfaces. *J. Graph Theory* **2010**, *63*, 17–30. [CrossRef]

13. King, E.L.; Pelsmajer, M.J. Dominating sets in plane triangulations. *Discret. Math.* **2010**, *310*, 2221–2230. [CrossRef]
14. Ananchuen, N.; Ananchuen, W.; Plummer, M.D. *Domination in Graphs*; Birkhauser: Boston, MA, USA, 2011.
15. Campos, C.N.; Wakabayashi, Y. On dominating sets of maximal outerplanar graphs. *Discret. Appl. Math.* **2013**, *161*, 330–335. [CrossRef]
16. Li, Z.P.; Zhu, E.Q.; Shao, Z.H.; Xu, J. On dominating sets of maximal outerplanar and planar graphs. *Discret. Appl. Math.* **2016**, *198*, 164–169. [CrossRef]
17. Liu, C.J. A note on domination number in maximal outerplanar graphs. *Discret. Appl. Math.* **2021**, *293*, 90–94. [CrossRef]
18. Chellali, M.; Volkmann, L. Relations between the lower domination parameters and the chromatic number of a graph. *Discret. Math.* **2004**, *274*, 1–8. [CrossRef]
19. Hedetniemi, S.M.; Hedetniemi, S.T.; Laskar, R.; Mcrae, A.A.; Wallis, C.K. Dominator partitions of graphs. *J. Comb. Inf. Syst. Sci.* **2009**, *34*, 183–192.
20. Gera, R.M.; Rasmussen, C.; Horton, S. Dominator colorings and safe clique partitions. *Congr. Numer.* **2006**, *181*, 19–32.
21. Gera, R.M. On dominator colorings in graphs. *Graph Theory Notes N. Y.* **2007**, *52*, 25–30.
22. Gera, R.M. On the dominator colorings in bipartite graphs. In Proceedings of the 4th International Conference on Information Technology, New Generations, Las Vegas, NV, USA, 2–4 April 2007; pp. 947–952.
23. Kavitha, K.; David, N.G. Dominator coloring of some classes of graphs. *Int. J. Math. Arch.* **2012**, *3*, 3954–3957.
24. Chellali, M.; Maffray, F. Dominator colorings in some classes of graphs. *Graphs Comb.* **2012**, *28*, 97–107. [CrossRef]
25. Merouane, H.B.; Chellali, M. On the dominator colorings in trees. *Discuss. Math. Graph Theory* **2012**, *32*, 677–683.
26. Arumugam, S.; Bagga, J.; Chandrasekar, K.R. On dominator colorings in graphs. *Proc. Math. Sci.* **2012**, *122*, 561–571. [CrossRef]
27. Kazemi, A.P. Total dominator chromatic number of a graph. *Trans. Comb.* **2015**, *4*, 57–68.
28. Kazemi, A.P. Total dominator coloring in product graphs. *Util. Math.* **2014**, *94*, 329–345.
29. Kazemi, A.P. Total dominator chromatic number and Mycieleskian graphs. *Util. Math.* **2017**, *103*, 129–137.
30. Henning, M.A. Total dominator colorings and total domination in graphs. *Graphs Comb.* **2015**, *31*, 953–974. [CrossRef]
31. Merouane, H.B.; Haddad, M.; Chellali, M.; Kheddouci, H. Dominated colorings of graphs. *Graphs Comb.* **2015**, *31*, 713–727. [CrossRef]
32. Guillaume, B.; Houcine, B.M.; Mohammed, H.; Hamamache, K. On some domination colorings of graphs. *Discret. Appl. Math.* **2017**, *230*, 34–50.
33. Choopani, F.; Jafarzadeh, A.; Erfanian, A.; Mojdeh, D.A. On dominated coloring of graphs and some nordhaus–gaddum-type relations. *Turk. J. Math.* **2018**, *42*, 2148–2156. [CrossRef]
34. Krithika, R.; Rai, A.; Saurabh, S.; Tale, P. Parameterized and exact algorithms for class domination coloring. *Discret. Appl. Math.* **2021**, *291*, 286–299. [CrossRef]
35. Bondy, J.A.; Murty, U.S.R. *Graph Theory with Applications*; MacMillan: London, UK, 1976.
36. Klavar, S.; Tavakoli, M. Dominated and dominator colorings over (edge) corona and hierarchical products. *Appl. Math. Comput.* **2021**, *390*, 1–7. [CrossRef]

MDPI
St. Alban-Anlage 66
4052 Basel
Switzerland
Tel. +41 61 683 77 34
Fax +41 61 302 89 18
www.mdpi.com

Mathematics Editorial Office
E-mail: mathematics@mdpi.com
www.mdpi.com/journal/mathematics